David Page

Introductory Text-Book of Geology

David Page

Introductory Text-Book of Geology

ISBN/EAN: 9783742818003

Manufactured in Europe, USA, Canada, Australia, Japa

Cover: Foto ©Klaus-Uwe Gerhardt /pixelio.de

Manufactured and distributed by brebook publishing software
(www.brebook.com)

David Page

Introductory Text-Book of Geology

GEOLOGICAL EXAMINATOR.

Under this Title, and in compliance with frequent requests from Teachers and Students, the Author has prepared for his Text-Books (Introductory and Advanced) three progressive sets of Examination, with references, in each case, to the paragraphs of the Text on which the question is founded. Adapted to three grades of proficiency, and arranged, each as far as it goes, so as to present a systematic epitome of Geology, these series of questions enable the teacher to frame his examinations with greater sequence and connection than his time will ordinarily permit, while to the Student they afford a ready means of testing his own progress and proficiency.

PRICE NINEPENCE.

INTRODUCTORY TEXT-BOOK

OF

GEOLOGY

BY

DAVID PAGE, LL.D. F.R.S.E. F.G.S.

AUTHOR OF 'ADVANCED TEXT-BOOK OF GEOLOGY,' 'HANDBOOK OF GEOLOGICAL TERMS AND GEOLOGY,' 'PAST
AND PRESENT LIFE OF THE GLOBE,' 'GEOLOGY FOR GENERAL READERS,' 'INTRODUCTORY
AND ADVANCED TEXT-BOOKS OF PHYSICAL GEOGRAPHY,' ETC.

EIGHTH AND ENLARGED EDITION

WILLIAM BLACKWOOD AND SONS
EDINBURGH AND LONDON
MDCCCLXIX

"The study of the structure of the Earth must tend to enlarge the mind of man, in seeing what is past, and foreseeing what must come to pass in the economy of nature; and here is a subject in which we find an extensive field for investigation, and for pleasant satisfaction."—HUTTON's *Theory of the Earth*, 1795.

PREFACE.

The object of this little Treatise is to furnish an elementary outline of the science of Geology. In its preparation the utmost care has been taken to present a simple but accurate view of the subject, to lead the learner from things familiar to facts less obvious, and from a knowledge of facts to the consideration of the laws by which they are governed. By adopting such a method, Geology, instead of being a dry accumulation of facts, and its study a mere task of memory, becomes one of the most attractive departments of Natural Science, and affords one of the finest fields for the exercise of the observing and reflective faculties. The treatise, though initiatory, is arranged on a strictly scientific basis, the Author being convinced that the student's progress is best promoted by commencing at once with the technical treatment of his subject, and making him feel that he is step by step acquiring the power to master for himself the higher and more difficult deductions. Such a course may require closer attention, and cost him a little more labour at the outset; but it will be found, as he advances, to be the more pleasant as well as more profitable mode of procedure. Every science, like every manual art, has a style and mode of handling peculiar to itself—a fact too often lost sight of in volumes professing to be "Popular Treatises" and "Easy Introductions." A soldier does not acquire the ready use of his musket by being trained to handle a broomstick; neither can a student become familiar with the truths of his science, or be taught to apply them, through the medium of the language and illustration appropriate to the subject. Whatever may be the defects of this manual, the Author has endeavoured to write as a geologist—to afford the pupil an accurate outline of the science, should he stop short at this stage of his progress, and to present him, should he wish to prosecute the study, with a gradual introduction to a more advanced and comprehensive text-book. In either case the treatise is complete as far as it goes; and the student who has mastered its details will have acquired no insignificant amount of geological information.

GILMORE PLACE, EDINBURGH,
August 1854.

SECOND EDITION *published August* 1855.
THIRD EDITION *published April* 1857.
FOURTH EDITION *published April* 1860.
FIFTH EDITION *published November* 1861.
SIXTH EDITION *published November* 1864.
SEVENTH EDITION *published June* 1867.

EIGHTH EDITION.

THE continued demand for this little manual enables the Author, at frequent intervals, to make such alterations and additions as the progress of the science requires. While this revision, therefore, has been enlarged so as to embody the latest views and discoveries—as far as these are compatible with the nature of an elementary outline—the original textual arrangement has been carefully preserved, that as little inconvenience as possible may be felt in using it in the same class along with any of the later editions.

February 1869.

CONTENTS.

	PAGE
I. GEOLOGY: A HISTORY OF THE STRUCTURE AND CONDITIONS OF THE GLOBE, AS MANIFESTED IN ITS CRUST, OR PORTION ACCESSIBLE TO HUMAN INVESTIGATION,	9
II. CAUSES OPERATING ON THE CRUST OF THE GLOBE, AND MODIFYING ITS STRUCTURE AND CONDITIONS,	16
III. GENERAL ARRANGEMENT, STRUCTURE, AND COMPOSITION OF THE MATERIALS CONSTITUTING THE CRUST OF THE GLOBE,	29
IV. CLASSIFICATION OF THE MATERIALS COMPOSING THE EARTH'S CRUST INTO SYSTEMS, GROUPS, AND SERIES,	41
V. THE IGNEOUS ROCKS, AND THEIR RELATIONS TO THE STRATIFIED OR SEDIMENTARY SYSTEMS,	51
VI. METAMORPHIC OR NON-FOSSILIFEROUS SYSTEM, EMBRACING THE GNEISS, MICA-SCHIST, AND CLAY-SLATE GROUPS,	62
VII. THE LAURENTIAN AND CAMBRIAN SYSTEMS, EMBRACING THE EARLIEST FOSSILIFEROUS SCHISTS, SLATES, AND ALTERED LIMESTONES,	70
VIII. THE SILURIAN SYSTEM, EMBRACING THE LOWER AND UPPER SILURIAN GROUPS, OR THE LLANDEILO, WENLOCK, AND LUDLOW SERIES,	76

CONTENTS.

IX. THE OLD RED SANDSTONE OR DEVONIAN SYSTEM, EMBRACING THE GREY FLAGSTONE, THE RED CONGLOMERATE, AND THE YELLOW SANDSTONE GROUPS, ... 85

X. THE CARBONIFEROUS SYSTEM, EMBRACING THE LOWER COAL-MEASURES, MOUNTAIN LIMESTONE, AND TRUE COAL-MEASURES, ... 96

XI. THE PERMIAN SYSTEM, EMBRACING THE MAGNESIAN LIMESTONE AND LOWER NEW RED SANDSTONE, ... 111

XII. THE TRIASSIC SYSTEM, COMPRISING THE KEUPER, MUSCHEL-KALK, AND BUNTER SANDSTEIN OF GERMANY, OR UPPER NEW RED SANDSTONE OF ENGLAND, ... 118

XIII. THE OOLITIC SYSTEM, COMPRISING THE LIAS, THE OOLITE, AND WEALDEN GROUPS, ... 126

XIV. THE CHALK OR CRETACEOUS SYSTEM, COMPRISING THE GREEN-SAND AND CHALK GROUPS, ... 140

XV. THE TERTIARY SYSTEM, EMBRACING THE EOCENE, MIOCENE, PLIOCENE, AND PLEISTOCENE GROUPS, . . . 149

XVI. POST-TERTIARY SYSTEM, COMPRISING ALL ALLUVIAL DEPOSITS, PEAT-MOSSES, CORAL-REEFS, RAISED BEACHES, AND OTHER RECENT ACCUMULATIONS, ... 166

XVII. REVIEW OF THE STRATIFIED SYSTEMS.—GENERAL DEDUCTIONS, 183

GLOSSARIAL INDEX, ... 187

GEOLOGY.

I.

GEOLOGY: A HISTORY OF THE STRUCTURE AND CONDITIONS OF THE GLOBE AS MANIFESTED IN ITS CRUST OR PORTION ACCESSIBLE TO HUMAN INVESTIGATION.

Objects and Scope of the Science.

1. GEOLOGY (from two Greek words—*ge*, the earth, and *logos*, a discourse or reasoning) embraces, in its widest sense, all that can be known of the constitution and history of our globe. Its object is to examine the various materials of which our planet is composed, to describe their appearance and relative positions, to investigate their nature and mode of formation, the changes they have undergone and are still undergoing, and generally to discover the laws which seem to determine their characters and arrangements.

2. As a department of natural science, Geology confines itself more especially to a consideration of the mineral or rocky constituents of the earth, and leaves its surface-configuration to Geography, its vegetable life to Botany, its animal life to Zoology, and the elementary constitution of all bodies to the science of Chemistry. Being unable to penetrate beyond a few thousand feet into the solid substance of the globe, the labours of geologists are necessarily confined to its exterior shell or crust; hence they speak of the "crust of the globe," meaning thereby that portion of the rocky structure accessible to human investigation.

3. The materials composing this crust are rocks or minerals of various kinds—as granite, basalt, roofing-slate, sandstone, marble,

coal, chalk, clay, and sand—some hard and compact, others soft and incohering. These substances do not occur indiscriminately in every part of the world, nor, when found, do they always appear in the same position. Granite, for example, may exist in one district of a country, marble in another, coal in a third, and chalk in a fourth. Some of these rocks occur in regular layers or courses, termed *strata* (from the Latin word *stratum*, strewn or spread out), while others rise up in irregular mountain-masses. It is evident that substances differing so widely in composition and structure must have been formed under different circumstances, and by different causes; and it becomes the task of the geologist to discover those causes, and thus infer the general conditions of the regions in which, and of the periods when, such rock-substances were produced.

4. When we sink a well, for example, and dig through certain clays, sands, and gravels, and find them succeeding each other in layers, we are instantly reminded of the operations of water seeing it is only by such agency that accumulations of clay, sand, and gravel are formed at the present day. We are thus led to inquire as to the origin of the materials through which we dig, and to discover whether they were originally deposited in river-courses, in lakes, in estuaries, or along the sea-shore. In our investigation we may also detect shells, bones, and fragments of plants imbedded in the clays and sands; and thus we have a further clue to the history of the strata through which we pass, according as the shells and bones are the remains of animals that lived in fresh-water lakes and rivers, or inhabited the waters of the ocean. Again, in making a railway-cutting, excavating a tunnel, or sinking a coal-pit, we may pass through many successions of strata—such as clay, sandstone, coal, ironstone, limestone, and the like; and each succession of strata may contain the remains or impressions of different plants and animals. Such differences can only be accounted for by supposing each stratum or set of strata to have been formed by different agencies, under different arrangements of sea and land, as well as under different conditions of climate; just as at the present day the rivers, estuaries, and seas of different countries are characterised by their own special accumulations, and by the imbedded remains of the plants and animals peculiar to these regions.

5. In making such investigations, the geologist is guided by his knowledge of what is now taking place on the surface of the globe—ascribing similar results to similar or analogous causes. Thus, in the present day, we see rivers carrying down sand and mud and gravel, and depositing them in layers, either in lakes,

OBJECTS AND SCOPE OF GEOLOGY.   11

in estuaries, or along the bottom of the ocean. By this process many lakes and estuaries have, within a comparatively recent period, been filled up and converted into dry land—the layers of sand and mud gradually consolidating and hardening into rocky strata. We see also the tides and waves wasting away the sea-cliffs in one district, and accumulating expanses of sand and salt-marsh in some sheltered locality. By these agencies thousands of acres of land have been washed away and covered by the sea, even within the memory of man; while by the same means new tracts have been formed in districts formerly covered by the tides and waves. Further, we learn that, during earthquake convulsions, large districts of country have sunk beneath the waters of the ocean; while in other regions the sea-bed has been elevated into dry land. Volcanic action is also sensibly affecting the surface of the globe—converting level tracts into mountain-ridges, throwing up new islands from the sea, and casting forth molten lava and other materials, which in time become hard and consolidated rock-masses like the greenstones and basalts of the older hills.

6. As these and other agents are at present modifying the surface of the globe, and changing the relative positions of sea and land, so in all time past have they exerted a similar influence, and have necessarily been the main agents employed in the formation of the rocky crust which it is the province of Geology to investigate. This world of ours is and has ever been subject to incessant *waste* and *reconstruction*—here wasted and worn down by frosts, rains, rivers, waves, and tides, and there built up again by the deposition of the water-borne materials, by the growth of plants and animals, and by the accumulation of volcanic ejections. Not a foot of the land we now inhabit but has been repeatedly under the ocean, and the bed of the ocean has formed as repeatedly the habitable dry land. No matter how far inland, or at what elevation above the sea, we now find accumulations of sand and gravel,—no matter at what depth we discover strata of sandstone or limestone,—we know, from their composition and arrangement, that they must have been formed under water, and brought together by the operations of water, just as layers of sand and gravel and mud are accumulated or deposited at the present day. And as earthquakes and volcanoes break up, elevate, and derange the present dry land — here sinking one portion, there tilting up another, and everywhere producing rents and fissures—so must the fractures, derangements, and upheavals among the strata of the rocky crust be ascribed to the operation of similar agents in remote and distant epochs.

7. By the study of existing operations, we thus get a clue to

the history of the globe ; and the task is rendered much more certain by an examination of the plants and animals found imbedded in the various strata. At present, shells, fishes, and other animals are buried in the mud or silt of lakes and estuaries ; rivers also carry down the remains of land animals, the trunks of trees, and other vegetable drift ; and earthquakes submerge plains and islands, with all their vegetable and animal inhabitants. These remains become enveloped in the layers of mud and sand and gravel formed by the waters, and in process of time are *petrified* (Lat. *petra*, a stone, and *fio*, I become); that is, are converted into stony matter like the shells and bones found in the deeper strata. Now, as at present, so in all former time must the remains of plants and animals have been similarly preserved ; and as one tribe of plants is peculiar to the dry plain, and another to the swampy morass,—as one family belongs to a temperate, and another to a tropical region,—so, from the character of the imbedded plants, are we enabled to arrive at some knowledge of the conditions under which they flourished. In the same manner with animals : each tribe has its locality assigned it by peculiarities of food, climate, and the like; each family has its own peculiar structure for running, flying, swimming, plant-eating or flesh-eating, as the case may be ; and by comparing *fossil* remains (fossil, from Lat. *fossus*, dug up, applied to all remains of plants and animals imbedded in the rocky crust) with existing races, we are enabled to determine many of the past conditions of the world with considerable certainty.

8. By examining, noting, and comparing as indicated in the preceding paragraphs, the geologist finds that the strata composing the earth's crust can be arranged in groups or series ; that one set or series always underlies, and is succeeded by another set ; and that each series contains the remains of plants and animals not to be found in any other series. Having ascertained the existence of such a sequence among the rocky strata, his next task is to determine that sequence in point of time—that is, to determine the older from the newer series of strata ; to ascertain, if possible, the nature of the plants and animals whose remains are imbedded in each set ; and, lastly, to discover the geographical range or extent of the successive series. These series he calls *formations*, as having been formed during different arrangements of sea and land, and under the varying influences of climate and other external conditions; and it is by a knowledge of these that the geologist is enabled to arrive at something like a history of the globe— imperfect, it may be, but still sufficient to show the numerous changes its surface has undergone, and the varied and wonderful

races of plants and animals by which it has been successively inhabited. To map out the various mutations of sea and land, from the present moment to the earliest time of which we have any traces in the rocky strata; to restore the forms of extinct plants and animals; to indicate their habits, the climate and conditions under which they grew and lived,—to do all this, and trace their connection up to existing races, would be the triumph, as it is now the aim, of all true geology.

Theoretical and Practical Bearings of the Science.

9. Such are the objects and scope of what may be termed THEORETICAL GEOLOGY—a science of comparatively recent growth, but of high and enduring interest. The problems it proposes to solve are among the most attractive that can engage the ingenuity of man—leading him from his own position and connection with the present aspects of this planet, back through all its former conditions, to the time when it came fresh and glowing from the hand of its Almighty Maker. As a legitimate cultivator of natural science, the geologist bases his deductions on numerous and well-observed facts; collects, arranges, and compares, with honest and scrupulous care; and by such means proceeds from phenomena that are obvious and taking place around him, to the explanation of those that are remote and less apparent. His object is to unfold the history of our globe as revealed in the composition and arrangement of the rocky crust, with all its imbedded remains of past life; and whether in collecting data among the hills and ravines, by the sea-cliff or in the mine, or in arranging and drawing from these data the warranted conclusions, the earnest student will find Geology at once one of the most healthful and exhilarating, as it is intellectually one of the most attractive and expanding, of human pursuits.

10. Nor is the science, in a PRACTICAL point of view, of less importance to man. Deriving, as we do, all our metallic and mineral stores from the crust of the earth, it is of vast utility to be able to distinguish correctly between mineral substances, to determine in what positions they occur, and to inform the miner with certainty where they are, or are not, to be found. Again, the engineer, in tunneling through hills, in cutting canals, excavating harbours, sinking wells, and the like, must, to do his work securely and with certainty, base in a great measure his calculations on the nature of the rocky materials to be passed through—information he can only obtain through the deductions of Geology. The architect

also, in selecting his material, by attending to the formation and texture of the rock, and observing how it has been affected by the weather in the cliffs and ravines, may often avoid the use of a wasting and worthless building-stone; while his knowledge of geological succession will enable him to detect in different localities the same material. The farmer, in like manner, whose soils are either formed by the disintegration of the subjacent rocks, or are affected by their retentive or absorbent nature, may learn much useful information from the demonstrations of the geologist. The study of physical geography, in relation to the migrations and habitats of plants and animals, the acclimatising and cultivation of certain animals and vegetables, and even touching the development and health of man himself, can only attain the character and position of a science, if treated in connection with the fundamental doctrines of geology. The artist and landscape-gardener may also reap substantial benefit from a study of the science as bearing on surface-configuration and character of scenery: and though such a knowledge, of itself, will make neither artists nor landscape-gardeners, it will often prevent them from committing unpardonable outrages on the landscapes of nature.

11. To acquire a knowledge of the science sufficient for the purposes indicated in the preceding paragraph, is not a very difficult task. The objects of research are scattered everywhere around us. Not a quarry by the wayside, not a railway-cutting through which we are carried, not a mountain-glen up which we climb, nor a sea-cliff under which we wander, but furnishes, when duly observed, important lessons in geology. A hammer to detach specimens, and a bag to carry them in, a sketch-book to note unusual appearances, an observing eye, and a pair of willing limbs, are nearly all the young student requires for the field; and by inspection and comparison in some museum, and the diligent use of his text-book, he will very shortly be able to proceed in the study as a practical observer. Let him note every new and strange appearance, handle and preserve every specimen with which he is not familiar—throwing nothing aside until he has become acquainted with its nature; and thus, besides obtaining additional knowledge and facilitating his progress, he will shortly acquire the invaluable power of prompt and accurate discrimination.

RECAPITULATION.

In the preceding paragraphs we have endeavoured to explain that the object of Geology is to investigate the structure of the earth, in as far as that structure is accessible to human investigation. Combining all we know of this rocky structure, from the top of the highest mountain to the bottom of the deepest mine, it forms but an insignificant film of the four thousand miles which lie between the surface and centre of the globe. This film or outer portion is spoken of as the "crust of the globe," in contradistinction to the interior portions, of which we can know nothing by direct observation. Thin as this crust may appear, it is nevertheless the theatre of extensive, diversified, and ceaseless changes. Every change arising from the violence of the earthquake and volcano, every modification resulting from the waters that cover or course its surface, every operation dependent on atmospheric agency, as well as all that appertains to the development of vegetable and animal life, is performed on or within this shell. It is thus at once the theatre of all geological change, and the index to all true geological history. By noting the composition of its rocks, their position and succession, the space over which they spread, and the fossils they contain, the geologist is enabled to indicate the condition and appearance of the world during former epochs —to speculate as to the former distributions of sea and land, the modifications of climate thereby occasioned, and the kind of vegetables and animals that successively peopled its surface. To arrive at a rational history of the successive phases of the globe, is the aim of theoretical geology; to discover and classify its mineral stores—to ascertain their position and determine their abundance, so as to make them available for the industrial purposes of life—is the task of the practical geologist. Combining its economic with its speculative bearings, Geology becomes a science of high and enduring interest, deserving the study of every cultivated mind, and the encouragement of every enlightened government.

II.

CAUSES OPERATING ON THE CRUST OF THE GLOBE, AND MODIFYING ITS STRUCTURE AND CONDITIONS.

12. THE aim of Geology, as stated in the preceding chapter, is to furnish a history of the structure and past conditions of the earth. Had the exterior crust been subject to no modifying causes, the world would have presented the same appearance now as at the time of its creation. The distribution of sea and land would have remained the same; there would have been the same surface-arrangement of hill, and valley, and plain; and the same unvarying aspects of vegetable and animal existence. Under such circumstances, Geology, instead of striving to present a consecutive history of change and progress, would have been limited to a mere description of permanently enduring appearances. The case, however, is widely different; from the moment the earth began to revolve round the sun and rotate on its own axis, there has been one continuous round of change and progression. Winds, frosts, and rains; springs, streams, and rivers; tides, waves, and currents; the shivering of the earthquake, and the upheaving of the volcano; the alternate growth and decay of plants and animals; and the universal operations of chemical agency,—are all continually tending to separate, to combine, and to rearrange the materials composing the crust of the earth.

13. In a comparatively fixed and stable region like our own, one is apt to underrate the effects of these modifying causes. We see from our infancy the same hills and valleys, the same fields and streams, and are apt to infer that little or no change is going forward. As we note more attentively, however, we begin to perceive that changes have taken place—are yearly, daily, and hourly taking place around us. We see the river deepening its channel, the tides and waves wearing away the sea-cliffs, the frosts and rains crumbling down the rocky surface, the estuary filling up with sand-banks, and the lake in which we laved our young limbs

becoming shallower, and a large portion of it transformed into a marsh, luxuriant with reeds and rushes. If all this has taken place during some twenty or thirty years, what, we naturally ask, may have taken place during centuries?—and what the amount of change, when centuries have been multiplied by centuries? Nay, more, if a few years can work such changes in a district of comparative rest and stability, what are we to expect over the whole surface of the globe, and especially in regions whose lakes are like our seas, and compared with whose rivers our streams are tiny threads of water—regions of extremes, where rains fall in torrents—where inundations deface, earthquakes submerge, and volcanoes elevate and give birth to new mountains? Extending his views in this manner, the attentive observer soon discovers that the crust of the earth, instead of being a thing of permanence and stability, is subject to incessant change; and as he carries his thoughts over the lapse of centuries, he can readily perceive how sea and land may have frequently changed places—how old mountain-ranges may have been wasted and worn down, and new ones accumulated—the beds of lakes become *alluvial* tracts (Lat. *ad*, to; *luo*, I wash—formed by the operations of water), and the sands and muds of former shores been converted into solid strata.

14. The agents which produce these changes being universal and incessant in their action, and the causes of all geological phenomena, it is necessary the student should have a clear understanding of their nature and mode of operation. Taking it for granted that he is acquainted with the spheroidal figure of the earth, its daily rotation on its own axis, and its annual revolution round the sun; with the nature of the atmosphere that envelops it, and thus forms the medium in which clouds, rains, winds, storms, and changes of temperature are elaborated, and through which the heat and light of the sun are diffused; with the distribution of the earth's surface into land and water; and with the general geographical features of the different continents,—their mountain-ranges, plains, and valleys, as well as with the general characters of the plants and animals that people their surface,— we shall proceed to explain more minutely the nature of the causes now modifying the crust of the earth.

15. These modifying causes may be arranged under the following heads:—The ATMOSPHERIC, or those operating through the medium of the atmosphere; the AQUEOUS, or those arising from the operations of water; the ORGANIC, or those depending on vegetable and animal growth; the CHEMICAL, or those resulting from the chemical action of substances on each other; and the

IGNEOUS, or such as manifest themselves in connection with some deep-seated source of heat in the interior of the globe. These causes, being dependent on the original constitution of our planet, are of course everywhere present, and in ceaseless operation; acting silently and almost imperceptibly in one region, and violently, and on a gigantic scale, in another; scarcely appreciable in their results at one period, and producing at another the most extensive alterations on the surface-configuration of the earth.

Atmospheric Agencies.

16. Of these modifying causes, the ATMOSPHERIC, though not the most powerful, are by far the most general in their operations. The atmosphere envelops the earth on every side; acts mechanically by its currents of wind, chemically by the gases of which it is composed, and vitally in its being indispensable to vegetable and animal life. Thus winds blow and drift about all loose material, carrying them away from one spot, and piling them up in another. In this way extensive tracts (sand-dunes or link-lands) are formed along the coasts of many countries by the inland drifting of the shore-sands; and the loose arid sands of the African and Asiatic deserts are carried forward year after year over new expanses. The gases of the atmosphere (oxygen, nitrogen, and carbonic acid), partly by their own nature, and partly by the moisture always less or more present in the air, exert a wasting or *weathering* influence on all rock-surfaces—softening, loosening, and crumbling them down, to be the more readily borne away by currents of wind and water. Frost is also a powerful atmospheric agent. The rain and moisture that enter into the fissures of cliffs, and between the particles of all rocky matter, are often frozen during winter, and in this state of ice expand and force apart these rocks and particles. When thaw comes, the loosened particles fall asunder, and thus, in all latitudes where frost occurs, vast waste is every winter effected—most noticeably, of course, in elevated tracts, and in arctic and antarctic regions. It is also by the action of frost that the avalanche, glacier, and iceberg are formed: the *avalanche* of snow and ice, which is launched from the mountain-side, carrying masses of rock and soil before it—the *glacier* or ice-river that forms in the mountain-glen above, and slowly grinds its way to the valley below—and the *iceberg*, detached from some arctic shore, that floats its burden of boulders and rock-debris to warmer latitudes, there to drop them on the bottom of the ocean. As the diffuser of heat and light and

moisture, the atmosphere exerts important influences on vegetables and animals, making the surface teem with life in one region, and rendering it a barren and a solitary waste in another. The electric discharge of the thunderstorm, which occasionally shivers cliffs, is also elaborated in the atmosphere; while, generally speaking, this aerial envelope is the grand medium in which vapours, rains, hail, snow, and other aqueous manifestations, are incessantly engendered.

Aqueous Agency.

17. The modifying causes arising from the operations of water are in like manner universal and incessant. The AQUEOUS AGENCY (Lat. *aqua*, water) manifests itself most prominently in the mechanical effects of rains, springs, streams, rivers, waves, tides, and oceanic currents. Every shower that falls exerts a degrading or wasting influence on rocks, soils, and all exposed surfaces. By entering the pores of rocks and soils, rain softens and loosens their cohesion, and thus renders them more easily acted upon by currents of wind or water. Every runnel after a heavy rainfall bears its muddy burden to the neighbouring stream; and slight as this may seem during a single season or for any one district, yet when calculated for centuries, and for every region of the globe, the waste through this cause must be enormous. Landfloods and *freshets*, which tear up the soil and river-banks, also arise from heavy rains and sudden meltings of snow; and the periodical rainfall of the tropics produces inundations and all their accompanying effects on the land-surface. Springs, which are discharges of water from the earth, act both chemically and mechanically on the crust. Hot springs, and those whose waters contain carbonic acid, act chemically, by dissolving the rocks through which they percolate in the crust of the earth; and when they come to the surface, they partly deposit the matter which they held in solution, and partly carry it onward and forward to the ocean. Such springs are known as *mineral springs*, and generally indicate the kind of rock or mineral through which they have passed. Thus some are saline, or contain salt; some chalybeate (*chalybs*, iron), or contain iron; some calcareous (*calx*, lime), or contain lime; some silicious (*silex*, flint), or contain flint; while others give off sulphureous vapours, and various mineral admixtures. Such springs act chemically in dissolving and re-depositing mineral matter; and they act mechanically, in wearing and transporting, like all running waters. Streams and rivers act chiefly in a mechanical way, and their influence de-

pends partly on the nature of the rocks over which they run, the rapidity of their currents, their size or volume of water, and the amount of rock-debris or grinding material which they carry along with them. If the rocks through which they pass be of a soft and crumbling nature, they soon cut out channels, and transport the water-worn material, as mud, sand, and gravel, to the lower level of some lake, or to the bed of the ocean. Their cutting as well as transporting power is greatly aided by the rapidity of their currents; hence the effect of mountain torrents compared with the quiet and sluggish flow of the lowland river. The effect of rivers on the crust is thus twofold—namely, to waste and wear down the higher lands, and to bear along the material, and deposit it in valleys, in lakes, or in the ocean, in layers of mud, sand, and gravel. By such deposits lakes are silted or filled up, and become alluvial valleys; estuaries converted into level plains; and even large tracts reclaimed from the sea. As with rivers so with waves, tides, and ocean-currents; they all waste and wear away the sea-cliffs in one region, and deposit the *degraded* material (Lat. *de*, down, and *gradus*, a step—worn down step by step) in the state of mud, sand, gravel, and shingle in some sheltered locality. The least observant must have noticed the waste now going forward along many parts of our own shores, whereby fields and villages have been carried away, and at the same time the formation of new land in other parts by the deposition of this wave-worn and tide-borne material.

18. By the operations of water, as described in the preceding paragraph, vast changes have been effected, and are still in process of being effected, on the surface of the globe. There is scarcely a country in the world which does not present numerous glens and ravines and river-channels, all cut through the solid strata by the action of water; hence known as *valleys of erosion* (*erosus*, gnawed or wasted away). The rocky matter thus ground down is borne away by every flood, in the state of mud, sand, gravel, and shingle; and when the water comes to rest, these fall to the bottom as *sediment* (*sedere*, to settle or sink down). Every person must have observed the rivers in his own district, how they become muddy and turbid during floods of rain, and how their swollen currents eat away the banks, deepen the channels, and sweep away the sand and gravel down to some lower level. And if, during this turbid state, he will have the curiosity to lift a gallon of the water, and allow it to settle, he will be astonished at the amount of sediment or solid matter that falls to the bottom. Now, let him multiply this gallon by the number of gallons daily carried down by the river, and this day by years and

centuries, and he will arrive at some faint idea of the quantity of matter worn from the land by rivers, and deposited by them in the ocean. In the same way as one river grinds and cuts for itself a channel, so does every stream and rill and current of water. The rain as it falls washes away what the winds and frosts have loosened; the rill takes it up, and, mingling it with its own burden, gives it to the stream; the stream takes it up and carries it to the river; and the river bears it to the ocean. Thus the whole surface of the globe is worn and grooved and channeled—the higher places being continually worn down, and the wasted material carried to a lower level. By this process lakes are silted up and become marshes or plains, and estuaries and shallow seas are converted into tracts of alluvial land. The *delta* of the Nile (so called from the Δ, or delta-like shape of the space enclosed by the two main mouths of that river), the sunderbunds or mud-islands of the Ganges, the swamps of the Mississippi, the Amazon, the Niger, and Irawaddy, are examples of such deposits on a large scale; but every stream and current of water, however insignificant, is less or more performing a similar operation. Such deposits, when examined, are found to consist of layers of mud, vegetable drift, clay, sand, and gravel, containing, in greater or less abundance, the remains of plants and animals peculiar to the country through which the river flowed. In this manner layers or strata of sedimentary matter are forming at the present moment, and in like manner must they have been formed during all past ages of the world.

19. The general tendency of aqueous agency, whether operating as rivers or as tides, waves, and ocean-currents, is to wear down the higher portions of the earth's crust, and transport the material as sediment to some lower level. This sedimentary matter being merely floated (or *mechanically* suspended, as it is termed, in contradistinction to a *chemical* solution) in the current, the moment the water assumes a state of quiescence it begins to fall to the bottom. The heavier bodies, as shingle and gravel, fall first, next the finer particles of sand, and ultimately the light flocculent mud or clay. In this way we can account for the gravelly beach of one district, the sandy shore of another, and the muddy bottom of a third. The clayey mud of the great Chinese rivers tinges the waters of the Yellow Sea for upwards of fifty miles off shore, thereby giving it a name, and rapidly converting it into a shallow basin; the turbid waters of the Ganges discolour for many leagues the Bay of Bengal; and the mud of the Amazon is observable many hundred miles out in the Atlantic. Thus, year after year, a portion of the Himalayan Mountains is deposited in the Bay of

Bengal, and the waste of the Andes strewed along the bottom of the Atlantic, there to be re-formed into new strata, and constitute, it may be, the material of future continents.

Organic Agencies.

20. The ORGANIC AGENTS tending to modify the crust of the globe are those depending on vegetable and animal life. The term organic (from the Greek *organon*, a member or instrument) is applied to plants and animals, as being supplied with certain organs or members for the purposes of nutrition and growth. Their structure is said to be *organic*, and they are termed organised bodies in contradistinction to minerals, which are *inorganic*, and whose increase takes place by external additions, and not through the instrumentality of any peculiar organs. The growth and decay of vegetables are yearly adding to the soil, at the same time that they protect its surface from the wasting action of rain, frost, and the like. Accumulations of vegetable growth form peat-mosses, jungle, cypress, and other swamps; and the spoils of forests and the vegetable drift of rivers form rafts (like those of the Mississippi)—all of which are adding to the solid matter of the globe. Coal, as will afterwards be seen, is but a mass of mineralised vegetation; and under favourable conditions, and in course of time, submerged peat-mosses, jungle-growths, forest-growths, and drifted rafts would form similarly mineralised deposits. As vegetable growth is specially influenced by heat, moisture, and conditions of climate, so in certain regions will the influence of vegetation, as a modifying cause, be more felt than in others. As familiar instances of vegetable agency, we may point to the peculiar plants that spring up on the newly-formed *sand-dunes* (downs or hillocks of blown sand) by the sea-shore, and protect the surface from being blown and scattered about by the winds; to the peat-bogs of Ireland, Scotland, Holland, and Canada, often extending over thousands of acres, and varying from ten to forty feet in thickness; to the pine-rafts yearly floated down by the Mississippi; to the cypress-swamps of America; to the "tarai" or jungle-swamps of India; and to the mangrove growth that binds and protects the mud-flats and islands of such deltas as those of the Ganges, Irawaddy, and Niger.

21. The manner in which animals tend to modify the crust of the globe, is chiefly by adding their waste secretions or coverings. It is true that the bones and other remains of the larger animals are often buried in the mud of lakes and estuaries—there in time

to form solid petrifactions; but such results are lithologically trifling (that is, in a rock-forming sense—*lithos*, a stone), compared with shell-beds, infusorial accumulations, and coral-reefs. Thus, gregarious shell-fish—as oysters, cockles, and mussels—live in beds of considerable thickness, and, if entombed among the silt of estuaries, will in time form beds of shelly limestone like those occurring in the solid crust of the earth. Masses also of drift-shells occur less or more along the sheltered recesses of every ocean, and these, in like manner, when covered up or consolidated, will be converted into limestones and marbles. The recent discoveries of the microscope have shown that many accumulations of whitish mud in lakes and estuaries, as well as certain deposits in bogs and valleys now silted up, are almost wholly composed of the silicious remains of *infusorial* organisms (so called from being abundantly found in putrid vegetable infusions). These flinty cases are of extreme minuteness; but being aggregated in countless myriads, they constitute thick layers, as in the estuary of the Elbe, in the plains of the Amazon, and in many of our own bogs; just as the mountain-meal (*bergmahl*) of the Swedes, the edible clay of the Indians, and the polishing-slate of Tripoli, are analogous deposits of older dates. As with the infusoria so with the *foraminifera* (*foramen*, an opening, from the punctures in their shields through which they protrude their feelers), whose minute many-celled cases constitute a large proportion of the calcareous muds of the ocean, spreading for hundreds of miles over the sea-bed, just as they form the main mass of the chalks and nummulitic limestones of earlier epochs. When treating of the older rock-formations, we shall see what an important part these minute organisms have played in adding to the solid masonry of the globe; and were the accumulations now taking place in our seas and lakes and rivers investigated with proper care, we should in all likelihood discover them still playing as important a part in the formation of rock-material.

22. By far the most notable, as it is undoubtedly the most wonderful, exhibition of animal agency—or rather of animal-chemical agency—is that of the coral zoophyta. Endowed with the power of secreting lime from the waters of the ocean, the coral animalcule rears its *polypidom*, or rocky structure (*polypus*, and *domus*, a house), in the warmer latitudes of every sea—and there constructs reefs and barriers round every island and shore where conditions of depth and current are favourable to its development. Many of these reefs extend for hundreds of leagues and are of vast thickness, reminding one of the strata of limestone belonging to the older formations. The true reef-building

zoophyte is apparently limited in its range of depth, operating only where perpetually covered by the tide, and downwards to eighteen or twenty fathoms. Within this range it is ceaselessly active,—elaborating lime from the ocean, and converting it into a home for itself and its myriad progeny. Let any one examine a branch of common madrepore coral, count the number of cells or pores in it, remember that each pore is the abode of an independent but united being, and then reflect on the thousands of miles of coral-reef now in process of formation, and he will be lost in wonder at the numerical exuberance of animal life. The coral-reef occurs in all stages of development, from the living and growing branch to a compact and solid aggregation of limestone, scarcely to be distinguished from some of the softer marbles. Partaking of the elevation or depression of the sea-bottom, and being subject to the influence of the waves and breakers, a coral-reef is not a mere narrow ledge composed of various beautifully-formed corals, but a barrier of limestone more or less compact, mingled with sand, shells, sponges, sea-urchins, and other marine *exuviæ* (Lat. cast-off crusts, shells, and similar remains, see Glossary), and often presenting a surface above the waves weathered and converted into soil capable of sustaining a scanty vegetation.

Chemical Agencies.

23. The modifying causes resulting from CHEMICAL ACTION are numerous and complicated. Thus, the accumulation of the coral-reef is partly a chemical process; the operations of all mineral springs are more or less chemical; and many of the phenomena connected with volcanoes and earthquakes may arise from a similar source. Laying aside, in the mean time, the changes taking place in the interior of the rocky crust, by which some strata are consolidated and hardened, others softened and dissolved away, metallic veins formed, and new compounds elaborated by the union of different substances, we shall confine our remarks to those chemical results which chiefly appear on the surface. The formation of the coral-reef, we have said, is partly a chemical process. The limy matter is no doubt secreted by the polype, but its subsequent consolidation into a compact rocky mass is the result of chemical action among the particles of carbonate of lime, of which it is almost wholly composed. The same sort of cohesion takes place among drifted shell-beds and calcareous sands, often rendering them as hard and compact as ordinary building-stone, and then known as *littoral* or *shore-formed concrete* (Lat. *litus*,

the shore). Deposits of limestone from what are termed calcareous or petrifying springs are strictly of chemical origin, as are also the *stalactites* (Gr. *stalagma*, a drop), arising from the dropping of calcareous water from the roofs of caverns, and the *stalagmites* which incrust their floors. In this way are formed porous calcareous tufa or calc-tuff, compact calc-sinter (Ger. *sintern*, to drop), the travertin of Italy, and other calcareous aggregations. As with lime so in like manner with flint or silex—many hot springs, like those of Iceland and the Azores, depositing silicious incrustations (*silicious-sinter*), or permeating loose material, and binding them together with a hard flinty cement. Clay or alumina, sulphur, and other mineral matters, are also deposited largely from springs, or arise as sublimations from fissures connected with volcanic action. Deposits of salt, natron, and the like, are also of chemical origin; and under the same head may be classed all asphaltic or bituminous exudations, like the pitch-lakes of Trinidad and Barbadoes, the tar-springs of Rangoon, and the oil-wells of North America.

Igneous or Volcanic Agency.

24. The last and most important of the modifying causes to be noticed, are those depending on IGNEOUS or VOLCANIC AGENCY (Lat. *ignis*, fire). The operation of water, whether in the form of rain, rivers, or waves, is to wear down the higher portions of the earth's crust, and transport them to lower localities—thus tending to reduce all to one smooth and uniform level. This equalising tendency of water is mainly counteracted by the operations of fire —the earthquake and volcano breaking up, elevating, and producing that diversity of surface so indispensable to variety in vegetable and animal life. These two forces—the aqueous and igneous—may be considered as antagonistic to each other, and to them may be ascribed the principal modifications that have taken, or are still taking, place in the crust of the globe. Igneous agency, as depending on some deep-seated source of heat with which we are but little acquainted, manifests itself in three grand ways—viz., in Volcanoes, in Earthquakes, and in gradual Crust-Motions.

25. The effect of vulcanic, or internal igneous force, is to elevate, either by simple expansion and upheaval of the crust, or by the repeated accumulations of matter ejected from the interior. We can easily conceive of large areas of the earth's crust being fractured and borne up by volcanic force from beneath, and in this way many of our mountain-chains and hill-ranges have primarily

originated. At certain places openings or *craters* occur (so called from their cup-like form—Gr. *krater*, a cup or bowl), and from these are ejected at intervals molten lava, fragments of rock, ashes, dust, hot mud, and various gaseous exhalations. Flowing from the crater over the surrounding country, the lava, after cooling, frequently forms thick layers of rocky matter, varying in compactness from hard basalt to open and porous pumice-stone. Ashes, dust, and volcanic mud accumulate in a similar manner, eruption after eruption adding to the height of the mountain, and ultimately giving to it a conical form. In this way have the cones of Etna, Vesuvius, and Hecla been formed, and in this way have eruptions modified the surrounding country, filling up valleys, creating crags and cliffs, enveloping fields, and burying cities, as in the case of Pompeii and Herculaneum. As with the active volcanoes of Europe, so with hundreds of others in various parts of the globe; and as these cones are now gradually rising and enlarging, so, looking at many of our older hills and mountain-ranges, we discover abundant proofs of a similar origin and mode of formation. As yet we have spoken of volcanoes as taking place only on land; but we have also evidence of their occurrence in the ocean, creating shoals and islands like many of those in the Pacific. In the one case, volcanoes are termed *sub-aerial*, in the other *sub-aqueous*. When taking place under water, the volcanic discharges of lava and ashes will be interstratified and mingled with the sedimentary matter of the ocean—an occurrence we shall afterwards find very common among the older rock-formations. Even when seated on land, they are generally in close proximity to the sea, and thus their ejections of lava, dust, and ashes falling into the water get overlaid by true sediments; and hence, also, the alternations of igneous and aqueous materials that occur so frequently in the crust of the globe.

26. Earthquakes, which are intimately associated with volcanoes, and but expressions of the same internal force, produce modifications of the earth's crust chiefly by fracture, subsidence, and elevation. During their convulsions the level plain may be thrown into abrupt heights, rent with chasms and ravines, or even be submerged beneath the ocean. Their general tendency is, therefore, like that of volcanoes, to diversify the surface of the globe, and to render irregular what aqueous agency is perpetually striving to render smooth and uniform. During violent convulsions, extensive alterations are sometimes produced on the face of a country; and of such we have frequent and abundant record within the historical era. Even within the present century, we know that a large tract at the mouth of the Indus was submerged,

while a new district was raised from beneath the ocean; that the coast of Chili for many leagues was permanently elevated from six to ten feet; that a similar and more recent upheaval took place in North Island, New Zealand; and that in the West India islands, harbours have been sunk, towns destroyed, and rivers changed from their former courses.

27. The gradual Crust-Motions (whether elevations or depressions) connected with igneous agency are less obvious than the volcano and earthquake, but not on that account the less important or general. At present it is known, from repeated observation, that the northern shores of Scandinavia are gradually rising above the waters; the shores of Siberia, as well as of all the islands within the Arctic Circle, are fringed with numerous recent terraces; and large tracts along the eastern and south-western coasts of South America exhibit similar uprises. Such uprises, not being very perceptible, are apt to be under-estimated, or even disregarded; but when we cast our eye along the shores of our own island, and discover various ancient beaches or shore-lines stretching along above the present sea-level, at elevations varying from ten, twenty, forty, and sixty, to one hundred feet and upwards, we are then prepared to admit that such gradual uprises must be extensively modifying the appearance and conditions of the globe. Nor is it uprise alone, but depressions arising from the same cause are also observable in many regions, as among the islands of the Pacific, the Atlantic seaboard of the southern states of North America, the west coast of Greenland, and the southern shores of Norway. We know that the generic distribution of plants and animals is governed, in a great measure, by altitude above the sea; and one can readily perceive how such gradual uprises and depressions of the land must be gradually changing the character and distribution of the life upon its surface. Nor is it terrestrial existence alone that is influenced by such crust-movements: the sea-bottom is partaking of the same upward and downward motions, and marine life is even more sensitive than terrestrial to changes of depth and sea-bottom.

RECAPITULATION.

In the preceding chapter we have given a general outline of the causes now tending to modify the crust of the earth; that is, of the principal agencies concerned in the production of all geological change. These, we have said, were the Atmospheric, the Aqueous, the Organic, the Chemical, and the Igneous or Volcanic. By one or other of these agencies, or by a combination of them, are all the changes now taking place on the globe effected; and as we are warranted in concluding that these agents have similarly operated through all previous time, so to them must be ascribed the formation and structure of the solid crust. Rains, winds, and frosts must have always weathered and worn down; springs, streams, and rivers must have always cut for themselves channels, and transported the eroded material to lakes and seas, there to be spread out in layers and strata; and in these accumulations must the remains of plants and animals have been entombed, some swept from the land, and others buried as they lived in the waters. In this way, and by calling in the aid of chemical and organic agency to explain the occurrence of certain mineral-deposits and accumulations of vegetable and animal growth, we can account for the formation of all rocks which occur in layers or strata. On the other hand, as volcanic agency now breaks up the crust of the earth, elevating some portions and submerging others, and anon casting forth, from rents and craters, masses of molten matter, and showers of dust and ashes, so in former times must the same agency have fractured and contorted the solid strata, and cast forth molten matter, which, when cooled down, would form rock-masses, in which no layers or lines of deposit could appear. Besides modifying the earth's crust by upheaval and disruption, volcanic agency also produces a peculiar class of rocks; and these are found abundantly in all regions, from the recent lavas of Etna and Vesuvius to the basalts, greenstones, and granites of our older hills. We have thus, on and within the globe, a variety of agents ceaselessly active, and ceaselessly productive of change. The result of their operations is, and has ever been, the production of new rocks and new rock-arrangements; and these we shall now proceed to consider.

III.

GENERAL ARRANGEMENT, STRUCTURE, AND COMPOSITION OF THE MATERIALS CONSTITUTING THE CRUST OF THE GLOBE.

28. LAYING aside all speculations as to the interior constitution of the globe, concerning which we can know nothing by actual observation, we are warranted in saying, that the external portion or crust, accessible to human research, is composed of a variety of solid substances known as *rocks*. No matter whether in the form of soft and yielding clay, of loose sand and gravel, of beds of chalk and sandstone, or of masses of granite—all are termed by geologists, *rocks* and *rock-formations*. And the reason is obvious; the sand and gravel of the sea-shore are but the comminuted fragments of the cliffs above and have necessarily the same composition, while the mud and clay of the deeper waters are merely still finer comminutions of the same rocky material. Of such substances is the crust of the earth composed, and in one form or other we pass through them wherever we go beneath the surface—whether tunnelling through the hills, or sinking coal-mines in the level valleys. How these rocks are arranged, and of what they are composed, are the subjects of our present inquiry.

Stratified or Sedimentary Arrangement.

29. Judging from the operations of the modifying causes explained in the preceding chapter, one would naturally infer that all matter deposited as sediment from water would be arranged in layers along the bottom. Fine mud and clay readily arrange themselves in this manner, and sand and gravel are also spread out in layers or beds more or less regular. In course of time a series of beds will thus be formed, lying one above another in somewhat parallel order, thicker, it may be, at one place than at another, but still preserving a marked horizontality, and showing

30 IGNEOUS OR UNSTRATIFIED ARRANGEMENT.

distinctly their lines of separation or deposit. Thus the miscellaneous *debris* (a convenient French term for all waste or worn material, wreck, or rubbish) borne down by a river will arrange itself in such layers along the bottom of a lake—the shingle and

Stratified Arrangement of Sediments

gravel falling first to the bottom, next the finer sand, and, lastly, the impalpable mud or clay, as represented in the preceding diagram. In course of time a series of layers will be formed, not perfectly parallel, one above another, like the leaves of a book, but still spread out in a flat or horizontal manner. One cannot look at the face of a quarry, or pass through a railway-cutting, without observing how very generally the rocks are arranged in beds and layers. These layers are technically known as *strata* (plural of *stratum*); hence all rocks arranged in layers—that is, arising from deposition or sediment in water—are termed *aqueous, sedimentary,* or *stratified.* In applying these terms the student will perceive that *aqueous* refers to the agency by which such rocks have been produced, *sedimentary* to the mode in which they have been formed, and *stratified* to the way in which they are arranged.

Igneous or Unstratified Arrangement.

30. On the other hand, when we examine the rocky matter ejected from volcanoes, we observe no such lines of deposit, and no such horizontality of arrangement. In general, they break through the stratified rocks, or spread over them in mountain-masses of no determinate form—here appearing as walls, filling up rents and chasms, there rising up in huge conical hills—and in another region flowing irregularly over the surface in streams of lava. When such rocks are quarried or cut through, they do not present a succession of layers or strata, but appear in *amorphous* masses; that is, masses of no regular or determinate form (Gr. *a,* without, and *morphe,* form or shape). Thus, in connection with the stratified rocks, they present something like the annexed appearance,—A A A being stratified or sedimentary rocks lying

bed above bed, and B B being the igneous, rising up through them in massive and irregular forms.

Stratified and Unstratified Rocks.

Referring to their origin, they are spoken of as *igneous*, and to the way in which they have been produced, *eruptive;* while in contradistinction to the stratified rocks, they are termed the *unstratified*. We have thus in the crust of the globe, two great divisions of rocks, the STRATIFIED and UNSTRATIFIED—the one depending on the operations of water, the other resulting from the operations of fire; and, as we shall afterwards see, to one or other of these divisions do all rock-formations belong, however much broken up, displaced, and contorted, or how great soever the changes that have subsequently taken place in their mineral composition.

Relative Positions of Rocks.

31. Having been spread or strewn over the bottom of seas and lakes in the state of sediment, the original position of the stratified rocks must have necessarily been less or more horizontal. A bed of mud, for example, may be thicker in one part than in another; or it may thin out and altogether disappear, its place being taken by a deposit of sand or gravel; but still its general disposition is flat or horizontal. The stratified rocks, when broken up by earthquakes and volcanoes, will lose this horizontality, and be thrown into positions more or less inclined and irregular. Nay, by the violent and repeated operation of volcanic forces they may be thrown on edge, may subside in basin-shaped troughs and hollows, or be bent and contorted in the most strange and fantastic manner. Such appearances are frequent in sea-cliffs, in the sides of ravines, in railway-cuttings, and in quarries; and geologists speak of such faces or exhibitions of strata as *sections*—that is, *cuttings through*—exhibiting the order of relation among the several strata. The following section, for instance, exhibits strata at A in a *horizontal*

Horizontal and Inclined Stratification.

position; at B in an *inclined* position; at C in a highly inclined position, or *on edge;* and at D, thrown or *tilted up*. The angle at which a stratum inclines to the horizon is called its *dip;* and strata are accordingly said to dip at an angle of ten, twenty, or thirty degrees, as the case may be. When an inclined stratum comes to the surface, as at E, its edge is called the *outcrop,* and the line of outcrop is termed its *strike* or stretch, from the German word *streichen,* to stretch or extend. Thus we speak of the strike of a stratum being from east to west, and its dip to the north or south—the dip being always at right angles to the strike, and *vice versa.* When we walk along the edge or outcrop of a stratum, we follow its strike; when the quarryman excavates it as it slopes downward into the crust, he follows its dip. When strata dip in opposite directions from a ridge or line of elevations, as at F, the axis is termed *anticlinal* or *saddleback* (Gr. *klino,* I

Unconformable, Bent and Contorted Strata.

bend; *anti,* in opposite directions); and when they dip towards a common line of depression, as at G, the axis is said to be *synclinal* (Gr. *syn,* together), and the depression so formed is spoken of as a *trough* or *basin.* When bent and twisted, as at H, they are termed *contorted;* and the frequent bendings are spoken of as *flexures.* When strata lie upon each other in parallel order, they are said to be *conformable;* but when one set reclines upon another at a different angle, as at K, they are termed *unconformable.* In the diagram, for example, the horizontal series at K are *unconformable* to, or rest *unconformably* on the upturned edges of, the inclined series beneath them. Occasionally, the strata of a district, though lying at different angles, may all slope the same way, and in such a case they are said to be *monoclinal*

Monoclinal Strata.

(*monos,* alone), or dipping in one main direction, as M M M. Not unfrequently they are found in dome-shaped positions, and sloping on every side from a common centre or apex; and then they are said to be *periclinal* (Gr. *peri,* all around), or dipping in every direction. When strata terminate abruptly in a bold bluff edge, they are said to form an *escarpment* (Fr. *escarpé,* steep), as at L; and such escarpments may either be the sides of hills, sea or river cliffs, or precipitous heights now far removed from the influence of water. Patches or masses of strata detached from the main body of the formation to which they belong are termed *outliers*, as at O O ; and such outliers are often widely separated from their

Escarpment—Outliers.

original connection. In all cases of this kind, whether the outlier be an island detached from the parent continent, or an isolated mound in a valley, its connection is traced and confirmed either by the mineral similarity and succession of its strata, or by the identity of its fossils with those contained in the main formation.

32. Rocks of igneous origin present themselves in the crust of the earth, either as *disrupting*, *interstratified*, or *overlying* masses. Thus when igneous matter forces its way through the stratified rocks, and fills up the rents and fissures so made, it is termed dis-

Overlying, Interstratified, and Disrupting Masses.

rupting, as at A ; when, having passed through the strata, it spreads over their surface in sheet-like masses, as at B B, it is then said to be overlying ; and when these discharges have taken place at the bottom of the sea, and have been in turn covered over by new deposits of sediment, they then appear as interstratified with the true sedimentary rocks, as at C C. Occasionally the interstratified matter appears to have been ejected in the state of dust and ashes, and to have subsided as sediment in the ocean, there to be covered up by true aqueous debris ; but in such cases, an examination of the particles of the rock will generally determine its igneous origin—these being sharp and crystalline, and not water-worn and rounded as in the true aqueous deposits. Where volcanic dust and mud have mingled themselves with the sedimentary matter of the ocean, and been subsequently consolidated into strata, it is often impossible to distinguish between such compounds and rocks of true aqueous origin ; and for all practical purposes they may be regarded as ordinary sedimentary rocks.

33. The fissures and fractures produced in the rocky crust by

volcanic agency are known by such terms as *faults, slips, hitches,* &c. ; and when filled up with injections or infiltrations of mineral matter, they are spoken of as *dykes, lodes,* and *veins.* In the annexed diagram, A represents a slip or hitch, where one portion of

Slip, Fault, Dyke, and Veins.

the strata seems simply to have slipped down, while another portion has been hitched up ; B represents a fault, where the strata are not only displaced, but thrown up at different angles ; C a dyke, where the fissure has been filled with igneous matter, in the form of a dyke or wall ; and D a suite of lodes or veins passing partly through unstratified and partly through stratified rocks. Into the origin of these displacements the student at this stage is not required to enter, further than to remember that the tendency of every earthquake and volcano is to rend and shiver the solid strata ; that where the shock is unaccompanied by discharges of igneous matter, the fissures will simply be slips and faults ; that where it is accompanied by igneous discharges, the molten matter will force its way through, and fill up the fissures, producing dykes ; and that where the rents are subsequently filled up by infiltrations of mineral and metallic matter, the result will be lodes and veins. The production of dykes may be the work of a day; the infiltration of veins may require the lapse of ages.

Structure and Texture of Rocks.

34. Having seen that the crust of the earth is composed of stratified and unstratified rocks, and that these have been frequently broken through and displaced by volcanic agency, the student should next acquaint himself with the structure and mechanical characteristics of the different rock-masses. The terms employed to designate the external structure are not very numerous, nor are they very difficult to be understood. Thus, speaking of stratified rocks, we employ the terms *stratum* and *bed,* when the deposit is of some thickness ; *layer* or *band,* when it is thin, and holds a subordinate place among other strata ; and *seam,* when a rock of a peculiar character occurs at intervals among a series of strata. For example, the miner speaks of a *seam of coal* occurring among strata of clay and sandstone, and

of a *layer of ironstone* occurring in a bed of shale. Though the terms *bed* and *seam* are thus loosely used by many geologists as synonymous with layer and stratum, bed ought to be applied only to the surface junction of two different strata, and seam to the line of separation between them. Thus the upper surface of a stratum may be smooth, or it may be rough and irregular, and the under surface of the stratum laid above it must partake of the smoothness or this irregularity: this is *bedding;* the line that marks this separation between two strata is the *seam.* When certain kinds of strata split up into thin plates or laminæ, they are said to be *slaty, flaggy, fissile,* and *shaly.* Roofing-slate, for instance, is fissile or slaty; some sandstones used for paving are spoken of as flags or flagstones. When igneous rocks appear in columns, like the basalts of Staffa and the Giant's Causeway, they are termed *columnar;* and when the columns are irregular and not very distinct, they are said to be *sub-columnar.* Certain greenstones (called whinstones in Scotland, and used for road-making) appear in large square-like blocks—a structure which is styled *tabular* or *cuboidal;* and when igneous rocks of this class break up in masses of no regular shape or form, they are termed *massive* or *amorphous.*

35. The internal texture of rocks is also designated by terms expressive of the appearances they present when broken up by the hammer. Thus a rock is said to be *granular* when made up of distinct grains or particles, like granite; *saccharoid* (like loaf-sugar) when the grains have a uniform crystalline aspect, as in many statuary marbles; *porous* when full of pores or of open texture like pumice-stone; *vesicular* or *cellular* when full of little cavities like certain kinds of lava; *fibrous* when the texture is composed of fibres like asbestos; and *acicular,* or needle-shaped (Lat. *acus,* a needle), when the fibres are distinct and pointed. *Conglomerates* are rocks composed of water-worn pebbles—in other words, consolidated gravel; and *breccias,* or rocks of *brecciated* structure, are those in which the fragments are sharp and angular, from the Italian *breccia,* a crumb or fragment. When a rock is easily broken or crumbled down, it is said to be *friable; earthy* when the texture is soft and dull; *compact* when of close and firm texture; *crystalline* when composed of sparkling or shining particles; and *sub-crystalline* when the lustre is somewhat dull and less apparent. The preceding are the terms most frequently employed by the geologist in describing rocks and rock-masses; the minuter distinctions of minerals, and the crystals of which they are composed, belong more especially to the study of Mineralogy.

Mineral Composition of Rocks.

36. The composition of the rocks constituting the crust of the globe may be viewed in two ways, either chemically or mineralogically. To the chemist every substance in nature is resolvable into certain primary elements; and of such elements upwards of sixty have been discovered—some gaseous, some liquid, some solid, some metallic, and others non-metallic. In examining a piece of marble, for example, the chemist resolves it into carbonic acid and lime; or, more minutely, into oxygen, carbon, and a metallic element called calcium. It is enough for the mineralogist, on the other hand, to know that it is a limestone, and to describe it as pure or impure, as soft or compact, as earthy or crystalline. The geologist, again, regards more especially its position and mode of occurrence, with what rocks it is associated, what fossils are imbedded in it; and from these and other data endeavours to arrive at the conditions under which it was formed, and the aspect of the world at the period of its formation. In drawing such conclusions, he is greatly aided by the deductions of chemistry and mineralogy; hence the importance of these sciences to the practical geologist. For elementary purposes it is enough to know the more familiar chemical substances and their leading compounds, such as the *gases*, oxygen, hydrogen, nitrogen, and chlorine; the *metals*, iron, gold, silver, copper, lead, zinc, tin, mercury, manganese, arsenic, and antimony; the *metallic bases*, sodium, potassium, aluminium, calcium, and magnesium, which, when united with oxygen, form the *earths* and *alkalies*—soda, potass, alumina or pure clay, lime, and magnesia; and the *non-metallic* bodies, silicon (*silex*, flint), carbon, sulphur, and phosphorus. For the same purpose the student should endeavour to make himself familiar with a few of the leading minerals and rocks; and this, in the absence of personal instruction in the field, he cannot more easily do than by the examination of properly-labelled specimens in a cabinet or museum. The following explanatory list may aid him at this stage of his studies:—

Sand—Gravel—Shingle.—Sand is an aggregation of water-worn particles, derived from rocks and solid substances; it is generally composed of quartz-grains, but may also be composed of particles of shells, corals, &c. Gravel is the term applied when the particles or fragments vary from the size of a pea to that of a hen's egg; and shingle when the fragments are larger and less rounded than those of gravel.

Sandstone—Grit—Conglomerate—Breccia.—Sandstone is simply consolidated sand, the particles having been compacted by pressure, or held together by lime, clay, oxide of iron, or some other cementing material. Grit is the term applied when the particles are hard and angular—that is, "sharper"—than in ordinary sandstones. Conglomerate, sometimes known as *puddingstone*, is a compact aggregation of gravel and pebbles of all sizes —in other words, consolidated gravel. Breccias, on the other hand, are rocks composed, as already explained in par. 35, of fragments more or less angular, and not water-worn and rounded as in conglomerates.

Mud—Clay—Silt.—These terms are applied to the fine impalpable particles of rock-matter worn and borne down by water, and deposited at the bottom of seas, lakes, and estuaries. Mud and silt, as composed of miscellaneous debris, are more friable than clay, which is plastic, and consists of silicious and aluminous particles.

Shale—Slate—Claystone.—Shale is merely consolidated mud and silt, assuming a structure less or more laminated or slaty. In slate the clayey particles predominate, and the consolidation is often so perfect that the rock assumes, as in ordinary roofing-slate, a semi-crystalline aspect. Claystone is massive and not slaty, and is of various origin. Clayey or argillaceous (Lat. *argilla*, clay) rocks, when breathed on, emit a peculiar and distinctive odour, which is easily recognised.

Marble—Limestone—Chalk—Marl.—The basis of all these rocks is carbonate of lime; that is, lime in chemical union with carbonic acid. The term marble is applied to the compact and crystalline varieties used for ornamental and statuary purposes; limestone, to the duller and less compact kinds used for mortar and in agriculture; and chalk, to the softer and earthier varieties. Marl is a loose term generally applied to friable compounds of lime and clay, and called clay-marl, or marl-clay, as the one or other ingredient predominates. As limestone is dissolved with violent effervescence by sulphuric and muriatic acids, its presence may be easily detected by a drop of either of these liquids.

Gypsum—Selenite—Dolomite.—Gypsum or stucco-rock is sulphate of lime, which, when calcined, forms the well-known plaster-of-Paris. Selenite is the term given to gypsum when crystallised and transparent. Dolomite (after the geologist Dolomieu) is a crystalline magnesian limestone—that is, a variety of limestone containing a large percentage of magnesia.

Quartz—Quartz-rock—Flint—Chert.—The basis of all these minerals is silex. Rock-crystal is pure crystallised silex; quartz is an impure silex, as in the clear hard crystals of granite, and the hard grains of sandstone. Quartz-rock is massive quartz occurring in veins or strata. Flint is impure nodules of silex; and chert, which is more opaque and less splintery, is an admixture of flint and lime.

Felspar—Felspar-rock—Porphyry.—The laminated glassy-looking crystals occurring in granite are of felspar. They can be scratched by a knife, while the quartz crystals resist it. Felspar-rock (or *felstone*) and felspar porphyry are amorphous rocks of felspar forming mountain-masses; and porphyry (Gr. *porphyreus*, purple, originally applied to a reddish-coloured igneous rock) may now be regarded as the general term for any igneous rock containing crystals distinct from the matrix or main substance in which they are imbedded.

Mica—Mica-schist—Gneiss.—The glistening, scaly, and transparent portions of granite are mica, so called from the Latin verb *mico*, I glisten.

It forms the principal ingredient in a set of slaty rocks called mica-schists; and it occurs in minute scales in many sandstones, giving to them a silvery appearance. Gneiss is also a laminated crystalline rock, differing from mica-schist (which is mainly composed of quartz and mica) in containing quartz, felspar, and mica, and in being more irregular and coarser in its lamination and texture.

Talc.—A transparent magnesian mineral resembling mica, but softer, and not elastic. It enters largely into the early slates, called talcose slates or talc-schists. When massive it forms talc-rock.

Hornblende—Augite—Hypersthene—Actynolite.—These are all dark, dark-green, or greenish-grey prismatic-like crystals, occurring largely in all the earlier and igneous rocks. They are closely allied in chemical composition, but differ in form and external aspect. Thus, hornblende (so called from its horny-like cleavage) generally occurs in larger crystals, is more silicious, and of less gravity than augite (*auge*, lustre); hypersthene is most frequently of a greenish hue, and in bladed crystals; while actynolite (Gr. *aktin*, a thorn) occurs in more slender and thorny-looking crystals. There are more minute and certain mineralogical differences, but at this stage the above may assist the beginner in his discriminations.

Granite—Syenite—Protogine.—Ordinary granite is composed of crystals of quartz, felspar, and mica, and appears in various shades of colour, as greyish, whitish, reddish, &c. When mica is wanting, and its place supplied by hornblende, the rock is called syenite, from Syene in Upper Egypt, where it occurs in abundance; and when talc takes the place of mica, the rock is known by the name of protogine. There are many varieties of granite, arising from such interchanges of minerals.

Basalt—Greenstone—Trap-rock—Trachyte—Lava.—These are rocks of igneous origin essentially composed of augite and felspar, with admixtures of hypersthene, hornblende, &c., and are chiefly distinguished by their hardness, compactness, and colour. Basalt is a close-grained, dark-coloured rock, often occurring in columns more or less regular; greenstone is not so close in the grain, is lighter in colour, and occurs either in tabular or amorphous masses. Trap-rock (so called from the step-like aspect it gives to hills composed of it) is a name which includes a great variety of igneous rocks, the general characters of which are easily recognised in the field. Basalt and greenstone may be included under the term trap, but the name is more generally applied to the looser and less crystallised masses known as trap-tuff, wacké, amygdaloid, &c. Trachyte is a trap or volcanic rock, so called from its rough, meagre feel (Gr. *trachys*, rough). Lava is the name given to all molten discharges from recent volcanoes.

Chlorite (from the Greek word *chloros*, greenish-yellow) is a mineral of a greenish-black colour, and generally of a foliated structure, in which condition it forms the principal ingredient in the greenish rock called chlorite slate or chlorite schist. Soft earthy varieties are known as *green-earth*.

Steatite—Soapstone—Serpentine.—All rocks containing steatite have a greasy or soapy feel, and are in this way easily distinguished. On this account some varieties have been termed soapstones. Serpentine, so called from its varying colours (like the back of a serpent), is one of the magnesian rocks occurring in primary districts, and often used for ornamental purposes. Chlorite is a common ingredient in these rocks.

RECAPITULATION.

Salts.—Salts of soda and potash, which occur in plains and deserts, are easily distinguished. Common salt, which is found in all sea-water, in many springs, and in vast masses, as rock-salt, is a chloride of sodium.

Bitumen is an inflammable mineral substance, found limpid as in naphtha, liquid as in rock-oil or petroleum, slaggy as in mineral pitch, and solid as in asphalt. It can likewise be distilled from coals and bituminous shales.

Coal—Anthracite—Jet—Lignite—Peat.—Coal is a well-known mineral, and may be briefly described as mineralised vegetable matter. It occurs in many varieties, as anthracite, which is non-bituminous, and burns without flame; caking-coal, cannel-coal, &c., which are all less or more bituminous, and consequently burn with smoke and flame. Jet is a compact lustrous variety of coal usually worked into ornaments; lignite (Lat. *lignum*, wood) or brown coal, is a variety of recent formation, in which the woody structure is distinctly apparent: and peat, which is still in growth, may be regarded as coal in its first stages of accumulation and mineralisation.

The Metals are either found *native*—that is, in a pure state—or combined with mineral matter, in the condition of *ores*. Gold, platinum, and silver are often found native in grains, scales, and nuggets; almost all the other metals occur in *ores*. Galena, for instance, is a sulphuret of lead; the little yellow cubes found in roofing-slate are iron pyrites, or sulphuret of iron; the brilliant green malachite, used for brooches, &c., is a carbonate of copper.

RECAPITULATION.

The object of the foregoing chapter has been to point out the arrangement, the structure, and composition of the rocks constituting the crust of the globe. Wherever a section of the crust has been exposed, whether naturally in cliffs and ravines, or artificially in quarries and mines, the rocks are found to be arranged either in layers or in indeterminate masses. Those arranged in layers have been evidently formed by the agency of water—those in shapeless masses by the agency of fire. The one set are termed the stratified, aqueous, or sedimentary—the other the unstratified, igneous, or eruptive. As a natural consequence of their origin, the igneous rocks break through, displace, and derange the original horizontal strata, which now appear inclined at various angles, fractured and contorted. The positions of the stratified rocks are indicated by such terms as plane, inclined, on edge, anticlinal, synclinical, bent, and contorted. Their dip or inclination is measured in reference to the horizon; their strike or line of outcrop is traced along the surface. The positions of the igneous rocks in reference to the stratified, are spoken of as disrupting, overlying, and interstratified; and the fractures or rents caused by volcanic convulsion, as fissures, faults, dykes, and veins. The structure and texture of rocks, whether of aqueous or of igneous origin, are distinguished by a variety of terms expressive of their appearance as they occur in the crust, or when broken up by the hammer. Thus the layers of the stratified rocks are spoken of as strata, beds, seams, bands, flags, slates, and schists, according to their thickness and mode of splitting; while the unstratified occur as columnar, sub-columnar, tabular, massive, and amorphous. As to the texture or internal structure of rocks, it is extremely varied, and is defined by such obvious terms as hard, compact, crystalline, saccharoid, granular, porous, vesicular, and the like. The composition of rocks, we have seen, may be viewed either in a chemical or mineralogical light; but, in whatever light they may be viewed, it is enough for the beginner in geology to be able to distinguish, at sight, such ordinary rocks as sandstone, conglomerate, shale, clay-slate, limestone, chalk, gypsum, coal, quartz, mica, felspar, granite, gneiss, greenstone, basalt, trap-tuff, lava, and a few of the ores of the more abundant metals. He should also early accustom himself to describe and group these rocks according to their most obvious or main ingredient, as—arenaceous or sandy (*arena*, sand); silicious or flinty (*silex*, flint); calcareous or limy (*calx*, lime); argillaceous or clayey (*argilla*, clay); carbonaceous or coaly (*carbo*, charcoal); and such like distinctions as will be clearly intelligible to others. Thus, we may have an argillaceous sandstone, a silicious limestone, and a calcareous shale; or, still more minutely, an argillo-calcareous sandstone, and a calcareo-arenaceous shale.

IV.

CLASSIFICATION OF THE MATERIALS COMPOSING THE EARTH'S CRUST INTO SYSTEMS, GROUPS, AND SERIES.

37. In order to arrive at a knowledge of the past conditions of the earth, it is necessary to examine not only the mineral character of the different strata, but to ascertain the nature of their imbedded fossils, their order of superposition, and other physical relations. By such an investigation we are enabled to determine their relative ages, to judge whether they were deposited in lakes, in estuaries, or in seas, and to say what kind of plants and animals flourished at the time of their formation. At the present day, the layers of mud, clay, sand, and gravel depositing in tropical estuaries and seas, will contain less or more the remains of plants and animals peculiar to the tropics: the deposits forming in temperate regions will contain, in like manner, the remains of plants and animals belonging to temperate climates; littoral or shore deposits (*littus, littoris*, the sea-shore) will contain shells differing from those that occur in the muds of the deeper ocean (*pelagic*); and should a time arrive when these layers are converted into solid strata, the fossilised plants and animals will become a certain index to the conditions of the regions and areas in which they were deposited. As with existing sediments, so with the strata constituting the solid crust: the lowest must have been formed first, the series beneath must be older than that above it; strata abounding in shells, corals, and other marine remains must have been deposited in the sea, while those containing fresh-water plants and animals give evidence of deposition in lakes or in estuaries. In this way we can determine the nature of any formation, and say whether it has been of *marine, lacustrine,* or *estuarine* origin. Again, igneous rocks, which displace and break through any set of strata, must be more recent than these strata; and if another set of strata overlie these igneous rocks, then must they have been deposited in water at a period subsequent to the

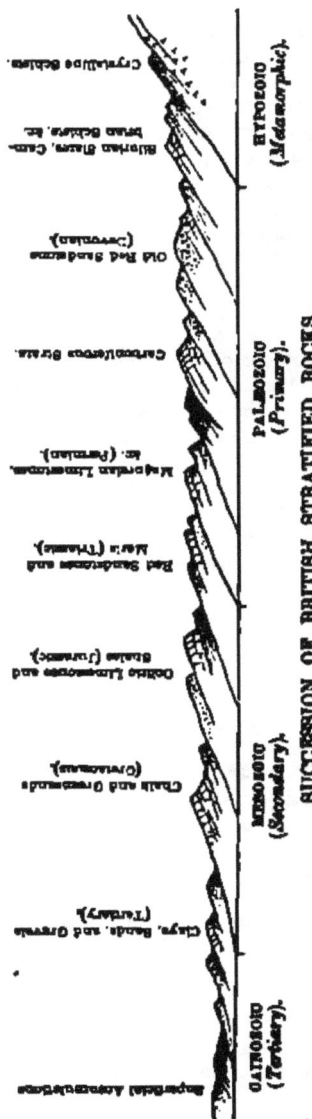

igneous eruptions. These and similar propositions are so apparent, that the student can have little difficulty in comprehending the principles upon which geologists have proceeded in classifying the rock-formations of the globe.

38. The principal guides to geological classification are, order of superposition among the strata, their mineral composition, and the nature of their imbedded fossils. The most superficial observer must have noticed the different aspects of the rocks in different districts, and a little closer inspection will enable him to detect that one set lies always beneath another set, and that while certain shells and corals are found in the lower series, the upper series may contain only the remains of terrestrial vegetation. Thus, in sinking a shaft in the neighbourhood of London, we would pass through thick beds of plastic clay, layers of sand, and strata of water-worn flint gravel; at Cambridge we would pass through strata of chalk; in the east of Yorkshire, through strata of fine-grained sandstones, and soft yellowish limestones called oolite; at Newcastle, through strata of shale, coal, and coarse-grained sandstones; in Forfarshire, through strata of red and greyish sandstones, and conglomerates; while on the flanks of the Grampians, we would pass through beds of roofing-slate and hard crystalline schists. On a minuter inspection of these strata, we would find that one series lay beneath or was older than another

series; that the chalk, for example, lay beneath the clays of London; that the yellow limestones of York lay beneath the chalk; that the coals of Newcastle were deeper seated than the oolites of York; and the red sandstones of Forfar still deeper than the coal-bearing strata. Further, when we began to examine the fossil contents of these different strata, we should find each set characterised by peculiar plants and animals—some containing marine shells and corals, some the remains of large reptiles and fishes, and others replete with the debris of terrestrial vegetation. By these methods we would soon be enabled to identify the chalk strata of Cambridge with those of Kent, the oolites of York with those of Bath, the coal-measures of Newcastle with those near Glasgow, and the slates of the Scottish Highlands with those of Cumberland and Wales. As with the rocks of Britain, so with those of every country investigated by geologists; and thus they have been enabled to arrive at a pretty accurate classification of the stratified rocks, both in point of time and mineral character.

39. About the beginning of the present century, the stratified rocks were classified as PRIMARY, TRANSITION, SECONDARY, TERTIARY, and RECENT; and though this arrangement has since been superseded by a more minute and perfect classification, the terms are still in daily use by geologists. Thus we hear and read of "secondary and tertiary rocks;" "rocks of the transition period;" and "fossils of the tertiary epoch." The terms, as well as the ideas they convey, are so firmly grafted on the language of geology, that it is necessary at this stage to present a tabular view of this arrangement:—

FORMATIONS.
{
RECENT.—All superficial accumulations, as sand, gravel, silt, marl, peat-moss, coral-reefs, &c. *Contain the remains of existing plants and animals only partially fossilised.*

TERTIARY.—Local and limited deposits of regular strata occurring above the chalk. *Contain the remains of plants and animals not differing widely in character from those now existing.*

SECONDARY.—Embracing all the strata known as chalk, oolite, lias, coal-measures, mountain limestone, and old red sandstone. *Contain fossil plants and animals of species altogether different from those now existing.*

TRANSITION.—Strata of slaty and silicious sandstones, calcareous shales, and limestones. *Contain few or no fossil plants, and the remains of no higher animals than crustacea, shell-fish, and zoophytes.*

PRIMARY.—All slaty and crystalline strata—as roofing-slate, mica-schist, and gneiss, very hard and compact, and totally destitute of organic remains.
}

40. By a more extensive examination of the strata in different countries, and especially by a more minute investigation of their fossil contents, these formations of the earlier geologists have since been subdivided into systems, groups, and series. This new arrangement has been founded either on mineral or on fossil distinctions—such differences being sufficient to warrant the conclusion that each set of strata was formed during successive epochs, and under different conditions of external nature. The terms formation, system, group, &c., are often loosely employed by geologists; but in the succeeding chapters we shall use the term *system* to signify any great assemblage of strata that have a number of mineral and fossil characters in common; the term *group*, to denote any portion of a system marked by a closer resemblance of mineral and fossil character; the term *series*, to designate any portion of a group which has some very marked character, either mineral or fossil; and so on with other subdivisions of the stratified formations. A system may thus comprehend several groups, a group several series, and a series may have several distinct stages at which some peculiar forms of life appeared in greatest abundance. Proceeding upon this principle, the stratified rocks may be subdivided into the following systems and groups, which derive their names partly from *lithological* considerations, as Chalk and Old Red Sandstone; partly from *geographical*, as Permian and Cambrian; and partly from *chronological*, as Tertiary and Post-Tertiary:—

I.—POST-TERTIARY or QUARTERNARY SYSTEM, comprising all alluvial deposits, peat-mosses, coral-reefs, raised beaches, and other recent accumulations. *Remains of plants and animals belonging to species now existing or but recently, and it may be only locally, extinct.*

II.—TERTIARY SYSTEM, embracing the "Drift," and all the regularly stratified clays, marls, limestones, and lignites, above the Chalk; arranged into pleistocene, pliocene, miocene, and eocene groups. *Remains of plants and animals for the most part extinct, but not differing widely from existing species.*

III.—CHALK or CRETACEOUS SYSTEM, embracing the chalk and greensand groups. *Remains of plants and animals chiefly marine, and belonging to species now extinct.*

IV.—OOLITIC SYSTEM, comprising the wealden strata, the upper and lower oolite, and the lias. *Remains of plants and animals (the most remarkable being huge reptilia), belonging to genera now extinct.*

V.—TRIASSIC SYSTEM, embracing the upper portion—saliferous marls, muschelkalk, and variegated sandstones—of what was formerly termed the "new red sandstone." *Remains of plants and animals more closely allied to those of the oolitic system above than to those of the carboniferous.*

STRATIFIED SYSTEMS. 45

VI.—PERMIAN SYSTEM, embracing the lower portion—magnesian limestones and red sandstones—of what was formerly termed the "new red sandstone." *Remains of plants and animals very closely allied, and often generically the same as those of the carboniferous strata.*

VII.—CARBONIFEROUS SYSTEM, embracing the coal-measures, the mountain limestone, and the carboniferous shales. *Remains of plants and animals abundant—the distinguishing features being an excess of endogenous vegetation in the coal-measures, and marine shells, fishes, and zoophytes in the mountain limestone.*

VIII.—OLD RED SANDSTONE or DEVONIAN SYSTEM, embracing the yellow sandstone, red conglomerate, and grey flagstone groups. *Remains of fishes, crustacea, shell-fish, &c., by no means rare; but terrestrial plants comparatively few, and imperfectly preserved.*

IX.—SILURIAN SYSTEM, embracing the upper and lower silurian groups, or the Ludlow, Wenlock, and Llandeilo series. *Remains of crustaceans, mollusca, radiata, zoophytes, and other marine invertebrate animals in abundance.*

X.—CAMBRIAN SYSTEM, embracing the upper and lower groups, and consisting of hard sub-crystalline slates and grits. *Remains of crustacea, mollusca, and zoophytes, with worm-burrows and other marine exuviæ.*

XI.—LAURENTIAN SYSTEM, consisting of crystalline schists, quartzites, and serpentinous limestones. *Remains of foraminiferal and other lowly and obscure organisms.*

XII.—METAMORPHIC ROCKS, embracing the clay-slate, mica-schist, and gneiss groups. *All hard and crystalline rocks devoid of fossils.*

41. Such are the stratified rocks when arranged in systems and groups; and, so far as geologists have been enabled to discover, there is no deviation from this order of succession. It must not be supposed, however, that all these groups are found at any part of the crust, lying one above another like the coats of an onion; on the contrary, only one or two of the groups may be developed, and these very scantily, and not in immediate succession. All that is meant by order of succession among the stratified rocks is, that wherever two or more systems come together, they are never found out of place; that is, the chalk is never found beneath the oolite, oolite beneath the coal, or coal beneath the old red sandstone. In Fifeshire, for example, the carboniferous system immediately overlies the old red sandstone; in Durham, the new red sandstone overlies the coal; in Yorkshire, the oolite overlies the new red sandstone, and the chalk the oolite; in Kent, the tertiary strata overlie the chalk; and

thus, though we do not find every series at one and the same place, we always find them occurring in the order above described. The old red sandstone and silurian, for instance, might be absent, and the coal in this case might rest on the clay-slate; or the new red sandstone and oolite might be absent, and chalk might rest on the coal; or even all of these might be wanting, and chalk immediately overlie the clay-slate. Still, there would be no reversal—a higher system would be overlying a lower; and the inference to be deduced would simply be, that the region in which any set of rocks was wanting, had been dry land during the deposition of these strata. This order of succession, or *superposition*, as it has been termed, is the great key to the solution of all geological problems; and so soon as an observer has fixed one point in the series, he knows infallibly his position in the history of the crust, no matter in what region he may be placed, or what the distance from the scene of his former observations. In determining his position, mineral characteristics may sometimes fail him, and a sandstone of the oolite may scarcely be distinguishable from a sandstone of the coal-measures; but palæontological characteristics are so constant, that the moment he discovers a few fossils, he is at once enabled to pronounce whether he is on an oolitic or on a carboniferous district.

42. The constancy of fossil characteristics has suggested the classification of the sedimentary rocks into certain great divisions, according to the types of living beings that predominated at certain epochs; but as these divisions are not yet very clearly determined, it may be enough for the pupil at this stage to understand the application of the following terms:—

CAINOZOIC PERIOD (*Recent Life*),	{ Post-tertiary or current epoch. { Tertiary epoch.
MESOZOIC PERIOD (*Middle Life*),	{ Cretaceous epoch. { Oolitic epoch. { Triassic epoch.
PALÆOZOIC PERIOD (*Ancient Life*),	{ Permian epoch. { Carboniferous epoch. { Devonian epoch. { Silurian epoch. { Cambrian epoch. { Laurentian epoch.
AZOIC PERIOD (*Void of Life*),	{ Non-fossiliferous epoch, or { Metamorphic system.

Instead of this arrangement it has been proposed by some to substitute the following, as sufficiently distinctive and more philosophical:—

STRATIFIED SYSTEMS.

NEOZOIC PERIOD
(*New Life*),
- Post-tertiary or present epoch.
- Tertiary epoch.
- Cretaceous epoch.
- Oolitic epoch.
- Triassic epoch.

PALÆOZOIC PERIOD
(*Ancient Life*),
- Permian epoch.
- Carboniferous epoch.
- Devonian epoch.
- Silurian epoch.
- Cambrian epoch.
- Laurentian epoch.

HYPOZOIC PERIOD
(*Beneath Life*),
- Metamorphic rocks in which fossils have not yet been detected.

In either case, all that is meant by the arrangement in the mean time is, that during certain epochs and during the deposition of certain formations there was a typical resemblance among the beings then peopling the globe; that down to the chalk, fossil species closely resemble those now existing (Gr. *kainos*, recent; *zoe*, life); from the chalk to the Permian the departure from recent types was greater (*mesos*, middle); and that from the Permian downwards the species were altogether distinct from the recent, and different in a majority of instances from those of the middle period (*palaios*, ancient). The term Neozoic (*neos*, new) merely expresses the distinction in broader terms; while Hypozoic (*hypo*, under) implies only the subjacent position of the metamorphic rocks—leaving it to future research to determine whether they are absolutely void of fossils or not. Whatever view may be adopted, the student should remember—and he cannot be too early cautioned ever to bear in mind—that throughout the whole of creation there is only ONE SYSTEM, and that in time past, as in time present, every aspect of nature gives evidence only of ONE all-pervading, all-directing Mind. The matter of the universe may undergo change of place, appearance, and arrangement; still it is the same matter, subject to the same laws that have operated through all time. The plants and animals on this globe may assume different specific aspects at different epochs and under different conditions, still they are constructed on the same plan and principle, and the laws which influence their being now, are identical with those that have governed vitality since the dawn of creation. Without this uniformity of law, the study of nature would be impossible. There is only ONE GREAT SYSTEM in creation, and the periods and systems of the geologist must be regarded as mere provisional expedients towards the elucidation and comprehension of that system.

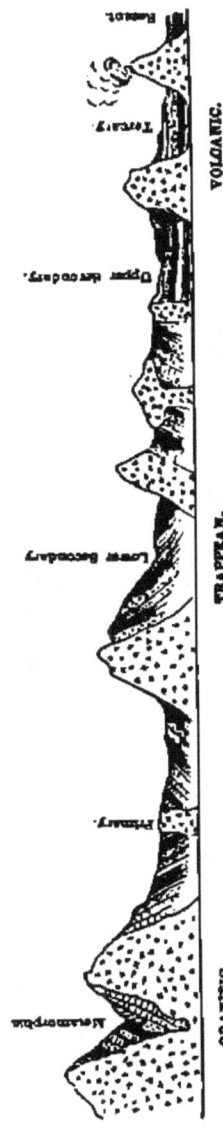

43. Besides these classifications of the stratified rocks according to their mineral characters, their fossil contents, and their order of superposition, there has also been attempted an arrangement of the unstratified or igneous masses. These, we have already seen, appear among the sedimentary strata without order or arrangement — heaving them out of their original horizontal positions, breaking through them in mountain-masses, or overspreading them after the manner of molten lava. Owing to this irregularity of origin, they are often better known by their mineral composition than by the order of occurrence. Still it is customary to speak of them as GRANITIC, TRAPPEAN, and VOLCANIC; meaning, by the term Granitic, the igneous rocks which, like granite, are usually found associated with the older strata; by the term Trappean, the igneous rocks most frequently associated with the secondary and tertiary strata; and by the term Volcanic, those that have made their appearance during the present epoch, as roughly sketched in the accompanying generalised section. It is true that it is often next to impossible to distinguish certain volcanic rocks from the more ancient traps; and it is also well known that granitic effusions occur among tertiary strata. Still, taking the three classes on the large scale, and looking at the stratified systems among which they usually occur, it will be found of essential service to retain the subjoined classification:—

VOLCANIC, { Lava, trachyte, scoriæ, &c., associated with recent accumulations.

TRAPPEAN,	{ Trap - tuff, amygdaloid, greenstone, basalt, &c., usually associated with tertiary and secondary strata.
GRANITIC,	{ Granite, syenite, porphyry, &c., usually associated with transition and primary strata.

RECAPITULATION.

The purpose of the preceding chapter has been to exhibit the classification adopted by geologists in describing the various rock-formations which constitute the crust of the globe. The bases upon which such a classification is founded are mineral composition, fossil contents, and order of superposition. By these aids the order of sequence among the stratified rocks has been pretty accurately ascertained; hence the subdivision of formations into systems, groups, and series. In making such an arrangement, it is not affirmed that any portion of the crust exhibits these systems one above another like the coats of an onion, but simply that one series always succeeds another in determinate order, and that though several series may be wanting in certain districts, such series as are present are never found out of their order of succession. Beginning at the surface, we have, in descending order—

1. Post-tertiary or recent accumulations.
2. Tertiary strata.
3. Cretaceous or chalk system.
4. Oolitic or Jurassic system.
5. Triassic, or upper new red sandstone.
6. Permian, or lower new red sandstone.
7. Carboniferous system.
8. Old red sandstone, or Devonian system.
9. Silurian system.
10. Cambrian system.
11. Laurentian system.
12. Metamorphic rocks.

The first eleven of these systems are spoken of as *Fossiliferous*, because they contain, less or more, the remains of plants and animals; the rocks of the last, which contain no traces of vegetable or animal life, are termed *Non-fossiliferous*. Referring to the fossil contents of the different strata, the term *Neozoic* (new life) is applied to the recent, tertiary, and upper secondary epochs; the term *Palæozoic* (ancient life) to the lower secondary and transition epochs; *Azoic* (or destitute of life) to the primary or non-fossiliferous epoch; or, avoiding all opinion as to the absence of fossils from these rocks, the term *Hypozoic* (beneath life) simply points out their position as lying under those systems that are decidedly fossiliferous. As with the stratified, so with the unstratified rocks: some acknowledged plan of classification is necessary, and that which arranges them into *Volcanic, Trappean*, and *Granitic*, is perhaps the most intelligible, as well as the most generally adopted. By employing the classification above indicated, every geologist, in treating of the rocks of a district, speaks a language intelligible to other geologists, and all the more intelligible that it is a classification founded on facts in nature, and not on mere arbitrary or technical distinctions.

V.

THE IGNEOUS ROCKS, AND THEIR RELATIONS TO THE STRATIFIED OR SEDIMENTARY SYSTEMS.

44. As previously stated, the igneous rocks have no determinate position in the crust of the earth. They derange, break through, and flow over the stratified formations; are of every age; and often of very complex mineral composition. With respect to their origin, geologists have as yet no satisfactory theory to offer, and generally content themselves by ascribing all igneous phenomena to some original source of heat existing within the globe. Various chemical hypotheses have from time to time been broached; and we know that the union of certain chemical substances takes place with a violent evolution of heat; but the occurrence of volcanoes, earthquakes, escapes of heated vapours and thermal springs, are by far too numerous and general to be accounted for on any principle of chemical union with which we are acquainted. Admitting the existence of some general and deep-seated source of heat to which all igneous rocks owe their origin, we shall proceed in the mean time to describe their characters and relations as classified under the heads GRANITIC, TRAPPEAN, and VOLCANIC.

Granitic Rocks.

45. *Granite* (*granum*, a grain) is named from its granular composition and aspect. The typical granite is a compound of quartz, felspar, and mica, arranged in distinct grains or crystals; and all rocks partaking of the character and appearance of granite are termed *granitic*. The granitic rocks are all highly crystalline; none of their crystals are rounded or water-worn; they present no traces of deposition or stratification; and they occur in the crust as mountain-masses and veins, bursting through and dis-

placing the sedimentary strata. From these circumstances, they are generally regarded as of igneous origin; though the character of their component crystals has induced many geologists to look upon them as highly metamorphosed rocks that have undergone change under great pressure and in the presence of water. Such a metamorphosis may be induced alike in igneous and in aqueous rocks; and in the present state of our knowledge, the general relations of the granites can be best explained by regarding them as deep-seated compounds of original igneous formation, but which have since been altered so as to assume their present highly crystalline character. As the earliest of igneous rocks, they are generally found associated with primary and transition strata, tilting them up on their edges, bursting through them in dykes and veins (*v v*), and variously altering their positions and mineral character.

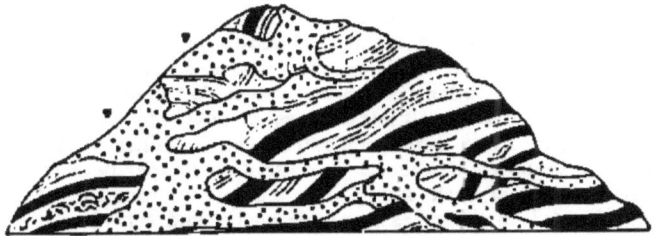

Granitic Veins (*v v*) traversing Gneiss at Cape Wrath.—M'Culloch.

Though occurring most abundantly among primitive strata, granitic outbursts may be found among rocks of all ages, but certainly not as a marked and general feature of the period.

46. Whether occurring in veins or mountain-masses, the structure of granite is irregular and amorphous. In its texture it varies from a close-grained compact rock to a coarse and loose aggregation of primary crystals. In the composition of granitic rocks there is also considerable variety, and the student will best learn to discriminate the different species by the examination of actual specimens. Ordinary *granite* is composed of crystals of felspar, quartz, and mica,—is of a greyish colour when the crystals of felspar are dusky white, and reddish when they are coloured by the presence of iron. When the dark glassy mineral called hornblende takes the place of the mica, the rock is known by the name of *syenite* (from Syene in Upper Egypt); and when talc supplants the mica, the admixture of felspar, quartz, and talc is known by the name of *protogine*. The term *hypersthenic granite* is applied to an admixture of quartz and

hypersthene, with scattered crystals of mica; and *porphyritic granite*, when, in addition to the crystals composing the general mass of the rock, there are indiscriminately mingled through it larger crystals of felspar, as in some of the granites of Devonshire. *Serpentine* is the name given to a compact admixture of variously-coloured minerals—as quartz, chlorite, steatite, &c., which produce a speckled and mottled appearance, resembling a serpent's skin; hence the term. Besides the above, there are other granitic compounds, in all of which quartz, felspar, mica, hornblende, and hypersthene are the principal ingredients, and talc, steatite, chloride, schorl, and actynolite, the accidental or modifying minerals.

47. Granitic rocks are widely distributed, and form the principal mass of many of the most extensive mountain-ranges in the world. The Grampians in Scotland, the mountains of Cumberland, Devonshire, and Cornwall in England, the Wicklow mountains in Ireland, the Dofrafelds in Scandinavia, the Alps in Switzerland, the Pyrenees in Spain, the Oural and Himalaya ranges in Asia, the Abyssinian and other chains in Africa, and the Andes in South America, are all less or more composed of granitic rocks, or of primary strata thrown up, and altered in mineral character by these granitic intrusions. Granitic districts, partly from the barren nature of their scanty soil, and partly from their high and elevated condition as mountain-chains and table-lands, are generally bleak and inhospitable, presenting few facilities for agricultural improvement or amenity.

48. The industrial purposes to which granitic rocks are applied, are alike numerous and important. As a durable building-stone for heavy structures, like docks, bridges, lighthouses, and fortresses, and as causeway-blocks for streets and highways, the harder varieties of granite are invaluable. As an ornamental stone for monuments, halls, chimney-slabs, pillars, pedestals, and the like, some varieties of granite (Peterhead, Cairngall, &c., in Aberdeenshire) are rapidly coming into use—the beauty and sparkle of their variegated texture, and the perfection to which they can be cut and polished, rendering their adoption peculiarly desirable. Some felspathic granites, like those of Devon and Cornwall, are easily decomposed when exposed to the weather, and in this state produce a fine impalpable clay known as *kaolin*, or china clay, and largely employed in the manufacture of the finest pottery and porcelain. *Apatite*, or crystallised phosphate of lime (Gr. *apaté*, deceptive), is another mineral product found in veins traversing the earlier igneous rocks, and promises to be of considerable value in the preparation of artificial manures.

Trappean Rocks.

49. The term *trap* (from the Swedish *trappa*, a stair, was originally applied to those igneous rocks which give to many hills of the secondary period a terraced or step-like appearance. Many of these rocks seem to have been formed under and in the proximity of water—here showered forth as volcanic dust and ashes, there as flows of lava, and anon interstratified with true sedimentary matter. It is to these successional flows of igneous matter, and the subsequent unequal degradation of the interstratified aqueous and softer rocks, that the trap hills owe their stair-like outlines. As the granitic rocks were generally associated with the older strata, so the trappean rocks are usually connected with the secondary, throwing them up on the sides of hills, breaking through them in dykes and veins, and spreading over them in sheet-like masses.

50. In their structure and composition, the trap rocks are extremely varied—some being compact and crystalline, as basalt and greenstone; others soft and earthy, as certain trap-tuffs, amygdaloids, and claystone-porphyries. Indeed, there is no class of rocks more puzzling either to the mineralogist or the geologist, their varieties being so numerous, and their relations to the strata

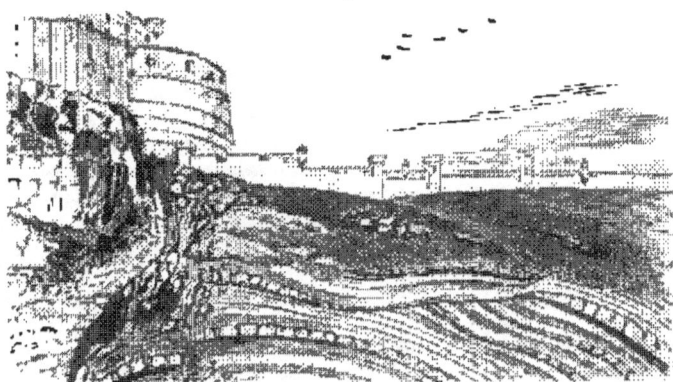

Basaltic Clinkstone of Edinburgh "Castle Rock," breaking through, contorting, and otherwise altering the stratified Shales and Sandstones of the Lower Coal-Measures.

being often so intricate and deceptive. The more crystalline varieties are known as basalts, greenstones, clinkstones, felstone and felstone-porphyries; the earthier varieties as trachytes, amygdaloids, trap-tuffs, and claystone-porphyries. Mineralogically speaking, they are chiefly composed of felspar, hornblende, and augite, with admixtures of hypersthene, olivine, green-earth, clay, and sulphuret of iron. The *basalts* are generally known by their columnar structure, by their dark close-grained texture, and by their enclosing spherical crystals of an olive-green mineral called olivine; the *greenstones* are less compact, more granular, exhibit distinctly the crystals of hornblende, hypersthene, &c., and often contain sulphuret of iron, and are usually massive or tabular in their structure; the *clinkstones* differ little from the greenstones in composition, but are more compact, break up into slaty fragments, and emit a ringing metallic sound when struck with the hammer; the *felstone rocks* are easily distinguished by their smooth compact texture and sub-crystalline aspect. The *amygdaloids* are rather earthy in texture, have been originally vesicular, and are so named from the almond-shaped concretions (Gr. *amygdalon*, an almond) of calc-spar, agate, and jasper, which now fill the vesicular cavities; the *trachytes* are greenish-grey varieties, indistinctly crystalline or earthy, and so named from the rough harsh feel (*trachys*, rough) they have to the finger; the *claystone-porphyries* are essentially of earthy felspar, with imbedded crystals of glassy felspar; and the *trap-tuffs* occur in every stage of texture, from soft scoriaceous masses to compact aggregations of rocky fragments, cemented together by igneous matter. Many of these rocks are evidently showers of volcanic dust and ashes that have fallen in the seas of deposit; and others are as evidently the broken and half-fused fragments of the associated strata. Much of the perplexing variety of texture that prevails among trap rocks seems to have arisen from the slowness or rapidity with which they have cooled; and we know, from actual experiments, that the same mass which will yield a compact basalt when suddenly cooled, will, when subjected to a slower process, produce a soft and earthy tufa.

51. The geographical area occupied by the trap rocks is very extensive, there being few secondary districts in which they do not rise up, either in undulating conical heights, or in terrace-like hill-ranges. Indeed, all the older secondary regions—that is, those occupied by the old red sandstone and carboniferous systems—owe their surface configuration chiefly to manifestations of trap. Much of this trap is of contemporaneous origin with the sedimentary rocks among which it occurs, and is of course interstratified with these deposits; but a great portion also is of posterior date, and in this case occurs as disrupting and overlying masses. To enumerate the districts in which trappean compounds occur, would be to map out the countries occupied with the whole transition,

secondary, and tertiary systems. In our own country, the Sidlaw, Ochil, Pentland, and Lammermuir ranges in Scotland; the Cheviot, Cumberland, Welsh, and Derbyshire hills in England; and most of the hills in Ireland, are of true trappean composition. The scenery produced by assemblages of trap hills is often extremely picturesque and beautiful; and the soil produced by their decomposition is generally so dry and productive, that the term "trap district" is usually regarded as synonymous with amenity and fertility.

52. The industrial purposes to which trap rocks are applied are numerous enough, but not of prime importance. Some basalts and greenstones make very durable building materials, but the difficulty of working such hard and refractory masses prevents their extensive use. The same may be said of the felspar-porphyries, clinkstones, and amygdaloids, which are rarely employed where the more easily dressed sandstones and limestones can be obtained. Their hardness, however, renders them peculiarly fitted for road material; hence their extensive use in causewaying and macadamising. From the *geodes* of the amygdaloids and trap-tuffs —that is, the crystals that coat or fill up the vesicular cavities in these rocks—are obtained most of the chalcedonies, agates, jaspers, and carnelians, made use of by the lapidary and jeweller.

Volcanic Rocks.

53. All the igneous rocks already described are, in one sense, *volcanic*—that is, have been produced by the agency of heat in a manner analogous to that of existing volcanoes. For the sake of classification, however, it is better to limit the term to such rocks as are now in process of formation, or have been formed since the close of the tertiary epoch. It may be difficult, in some instances, to distinguish a mass of trachytic lava from one of trachytic trap-tuff; but when the mass is viewed in connection with its associated rocks, its origin becomes readily apparent, and there is generally as little difficulty in distinguishing between recent volcanic products and trappean compounds, as there is in distinguishing between trap and granite. Volcanic rocks are therefore essentially products of the modern period, and are found, like the older igneous rocks, either elevating, bursting through, or overlying the stratified formations.

54. Volcanic products usually appear as lava, obsidian, pumice, scoriæ, ashes, hot mud, and various gaseous exhalations. *Lava* is the name given to the melted rock-matter ejected from active

craters, and which, when cooled down, forms varieties of volcanic tufa, trachyte, trachytic greenstone, and basalt, according to the varying proportions of felspar, hornblende, and augite in the mass, its fluidity when ejected from the crater, and the rapidity with which its cooling has been effected. *Obsidian*, or volcanic glass

View of Mount Etna—Conical Aspect of Volcanoes.

is a compact vitreous lava, in some instances scarcely to be distinguished from the slag of a glass-furnace. *Pumice* is a light porous rock, caused by the disengagement of gases in the mass while in a state of fusion; in other words, the solidified froth or scum of molten rock-matter. *Scoriæ, cinders, ashes*, and the like, are of the same mineral composition as the solidified lava, and seem to be produced by the explosive force of steam or other gases. *Volcanic mud*, which is found bubbling out from many fissures and hot springs, has a fœtid sulphurous odour, and in cooling and solidifying is often found to contain crystals of sulphur and gypsum. The *gaseous* products of volcanoes are for the most part sulphurous, and from this source is mainly derived the sulphur employed in the useful arts. All of these products are found less or more in every volcanic region; and the mode in which they are discharged, their varying admixtures, and the different appearances they assume, according to the rapidity or slowness with which they are cooled, afford highly instructive lessons to the geologist. Here the explosive force of highly heated vapours and molten matter breaks through and deranges the strata of the crust; there lava fills up the fissures, or, issuing from some vent, flows down the mountain-side, filling up valleys, damming up river-channels, and spreading over fertile plains: here scoriæ and ashes are showered forth, borne abroad by winds, and scattered over land and sea; there heated vapours are perpetually exhaling from rents and fissures, and incrusting their sides with mineral and metallic compounds. Discharge after discharge from volcanic vents gives rise, in course of time, to mountain masses; or, if spread along the bottom of the sea, is in turn overlaid by true sediment, and thus produces alternations of aqueous and igneous rocks. The molten matter also cools unequally—here forming a porous pumice, there a rough open tufa; here a granular trachyte, and there a compact mass, scarcely distinguishable from the older basalts and greenstones. And just as igneous forces are acting at the present day under the eye of the observer in the production of volcanic rocks, so must they have acted in former ages in the production of trappean and granitic compounds.

55. Although volcanic rocks are unknown in our own country, they occur extensively in many regions of the globe. In Europe there are three well-known centres of volcanic action—viz., that of Italy, to which Etna and Vesuvius belong; that of Iceland or Hecla; and that of the Azores. In Asia there are ample evidences of volcanic action along the borders of the Levant, the Caspian, and the Red Sea; in the islands (Mauritius, Comoro, &c.) of the Indian Ocean; throughout the whole of the Indian Archipelago; and northward through the Philippine, Japan, and Aleutian islands. In the Antarctic Ocean several cones of active eruption were discovered by our voyagers in 1841; and in the Pacific, the islands of New Zealand, the Sandwich, and other groups, are for the most part the results of volcanic action. In the Atlantic, the Canaries, Cape de Verd, Ascension, and other islands skirting the western coast of Africa, are well-known seats of volcanic action; while in the West Indies, and along the entire continent of America, from the islands of Tierra del Fuego (Land of Fire), northward through the Andes and Rocky Mountains, are numerous volcanic vents in a state of greater or less activity. In these centres and lines of igneous action (see accompanying Sketch-Map) many of the volcanoes seem to be *extinct;* some are merely smouldering or *dormant;* while others are incessantly *active,* either ejecting rocky matter from their craters, or rending the surrounding country by earthquake convulsions.

56. In an industrial point of view, volcanic products are of considerable importance. All, or nearly all, the *sulphur* of com-

VOLCANIC SERIES.   59

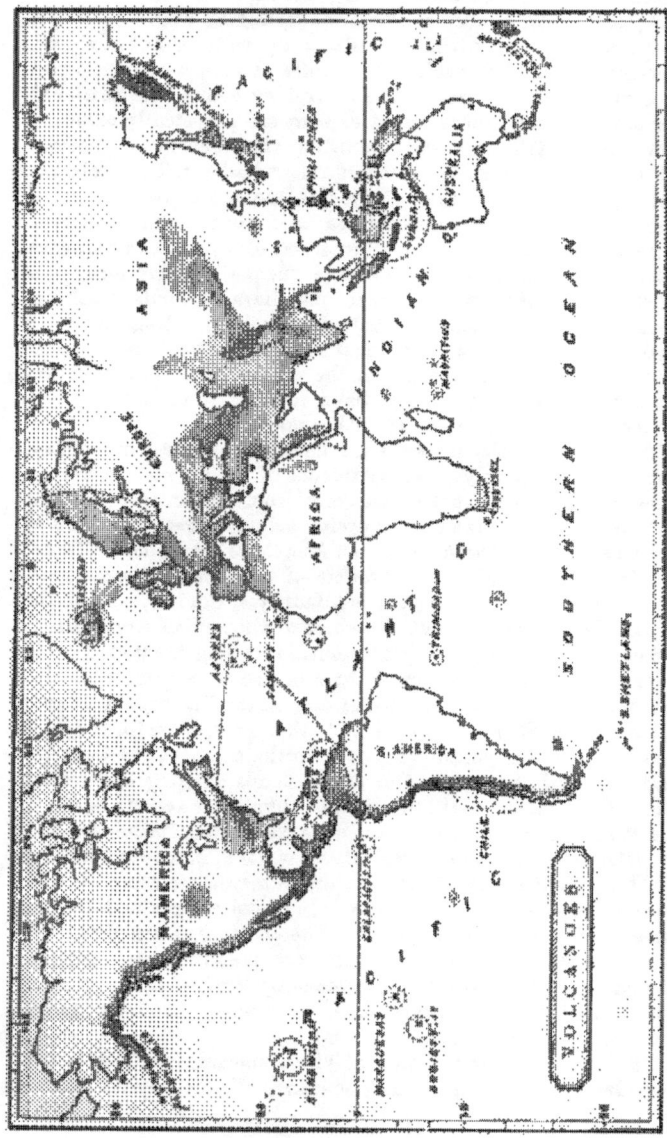

⁎ The shaded portions indicate the lines and craters of volcanic activity; the more important volcanoes being marked by a ⁎.

merce is derived from volcanic districts; and *sal-ammoniac* and *borax* have a similar thermal origin. Several of the *lavas* make a light and durable building-stone, and others are cut and polished for ornamental purposes like marble. *Pumice* has been long used as a polishing or rubbing stone; and *puzzolana*, which seems to be an altered felspathic tufa, is largely employed in the manufacture of Roman or hydraulic cement. *Obsidian*, as its name is thought to imply, was used by the ancients for mirrors; and the natives of various regions have used it, as our forefathers used flint, for knives, hatchets, and arrow-heads. Like the trap-rocks, many of the older lavas yield *agates, chalcedonies, olivine,* and other precious minerals.

RECAPITULATION.

The products described in the preceding chapter constitute one of the great divisions into which the rock-masses of the globe have been arranged. Though containing (unless in the case of dust and ashes that may have been showered abroad over the surrounding lands or into seas of deposit) no fossil record of the kind of plants and animals which have successively peopled the earth—and in this respect of less value in enabling us to decipher its history—they are still important monuments of past change; monuments in which we can trace the features of the world's former surface—its alternations of hill and valley, of sea and land, and of many of those external conditions which give character and colouring to organic life. Although ignorant of the origin of the internal heat or Vulcanism (*Vulcanus*, the god of fire) which gives rise to these products, we clearly perceive that its great world-function is to counteract, by upheaval from within, the waste and degradation to which the Earth's crust is exposed from without, and thus to preserve that variety of superficial conditions, so indispensable to the variety and wellbeing of vegetable and animal life. Viewed as the results of a great cosmical Law in diversifying the earth's surface, in shifting from area to area, and in obeying certain lines of direction, the igneous rocks assume a new interest; and it is only by studying their relations to, and the manner in which they have affected the stratified rocks, that we can ever hope to solve many of the most intricate problems in Geology. Their arrangement into GRANITIC, TRAPPEAN, and VOLCANIC, though partaking more of a mineralogical than of a geological distinction, is not without its value, so long as the student remembers that granite, though the deepest-seated igneous rock, may also be associated with strata of all ages, and that trap-rock, though most abundantly developed during the secondary period, may also be found in connection with strata of the earliest epochs. Bearing in mind these facts, and remembering also how similar many of these rocks are in mineral composition, and that they all occur in connection with the stratified formations—as *disrupting, overlying,* or *interstratified* masses—the student will readily perceive that it is chiefly in their mineral and mechanical bearings that he has to deal with them. Thus, the granitic masses are never scoriaceous and cellular, like recent volcanic rocks, nor are they ever earthy and amygdaloidal, like many of the trappean compounds. The trap group, as a whole, are less felspathic than the granites and porphyries, and exhibit a greater tendency to structural arrangement than either granitic or volcanic products; while the volcanic are decidedly more cellular and slag-like than either the granites or traps. Leaving special instances of igneous operation to be discussed in connection with the several stratified formations, we may briefly recapitulate that—

The VOLCANIC comprises lava, amorphous and basaltic, trachyte, obsidian, trachytic greenstone, tufa, pumice, scoriæ, ashes, &c.

The TRAPPEAN comprises basalt, greenstone, clinkstone, claystone, felstone-porphyry, amygdaloid, trap-tuff, tufaceous conglomerates, &c.

The GRANITIC comprises granite, syenite, protogine, serpentine, primitive greenstone, graphic granite, granitic porphyries, &c.

VI.

METAMORPHIC OR NON-FOSSILIFEROUS SYSTEM, EMBRACING THE GNEISS, MICA-SCHIST, AND CLAY-SLATE GROUPS.

57. WITHOUT adhering to the common belief, that granite forms the floor or basis on which all the stratified systems repose, it may at least be confidently asserted that granitic compounds upheave and break through the lowest known strata, and in this sense are certainly under-formed, or deeper seated than any other rocks as yet discovered by geology. Assuming, therefore, that the granitic group immediately underlies, and is intimately associated with, the lowest stratified rocks, we obtain a starting-point in the crust from which to commence an intelligible description of the systems that follow. The *Non-fossiliferous* system is so termed from its containing no remains of plants or animals, as far as geologists have been enabled to discover. For the same reason it has also been termed the *Azoic*, or destitute of life (*a*, without, and *zoe*, life), in contradistinction to the upper systems, which are all less or more fossiliferous. As this distinction, however, is founded solely on negative evidence, and as fossils may yet be discovered in some portion of these strata, it is thought better to employ the term *Hypozoic* (*hypo*, under, and *zoe*, life), which merely indicates that the system lies under all those that are known to be unmistakably fossiliferous. The name *Metamorphic* refers, on the other hand, to its mineral characteristics, and implies that the original structure and texture of its rocks have undergone some internal change or metamorphosis. At present these rocks are all less or more crystalline; their lines of stratification are often obliterated, or but faintly perceptible; and their whole aspect is very different from what is usually ascribed to rocks originally deposited in water. This change may have been brought about by the long-continued influences of heat and pressure, or it may be the result of some peculiar chemical action, combined with pressure, among the particles of which the rocks are composed. In whatever way

the metamorphosis has been effected, we see clearly that a change has taken place in the original sedimentary character of the strata, and that matter which at first consisted of water-worn debris—as silt, clay, and sand—has now been converted into hard, shining, and crystalline rocks. It must be remembered, however, that though mineral metamorphism is peculiarly the characteristic of the older and deeper-seated strata, it is by no means confined to gneiss and mica-schist; for, as we shall afterwards see, many sandstones of the later systems have been converted into quartzite or quartz-rock, many shales into jaspery hornstones, and many limestones and soft chalks into sparkling saccharoid marbles. In fact, wherever heat, chemical action, and pressure are present, there every species of rock will undergo a metamorphism or mineral change, and this in proportion to the length of time and the intensity of the operating agents to which it has been subjected. In this way much of the so-called *Metamorphic System* may yet be resolved into Laurentian, Cambrian, or Silurian strata, according as some less altered portions may be found to contain fossils that can be ascribed to one or other of these formations.

Gneiss and Mica-Schist Groups.

58. We arrange these groups under one head, for this reason, that though there is often a sufficient mineral distinction between gneiss and mica-schist when viewed on a large scale, there is, after all, very little difference in their geological character and history. In whatever state the particles of *Gneiss* may have been when originally deposited, we know that they now form a hard, tough, crystalline rock, exhibiting curved and flexured lines of stratification, and composed in the main of quartz, felspar,

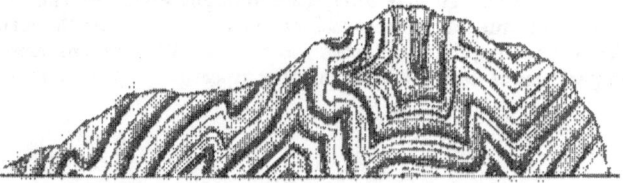

Fissured and Contorted Gneissose Rocks in Glen Quoich.—Murchison.

mica, and hornblende. Mineralogically speaking, it differs from the granitic rocks with which it is associated chiefly in this, that while the crystals of quartz, felspar, &c., are distinct and entire in granite, in gneiss they are indistinct, and confusedly aggregated. The main distinction, however, lies in the external character or structure of the two rocks—granite, like other igneous products, being strictly amorphous and homogeneous in the mass, gneiss exhibiting distinctly the lines and layers of deposit which evince its aqueous or sedimentary origin. What is thus affirmed of the sedimentary origin of gneiss is much more apparent in *Mica-schist*, which is often finely laminated and distinct in its lines of stratification. This distinction arises from its softer and less crystalline texture, as well as from the greater proportion of mica entering into its composition in the form of fragmented flakes or scales. There is, however, a great similarity between the two sets of rocks—beds and bands of gneiss interlacing and alternating with beds of mica-schist, and gneissose rocks frequently becoming so micaceous in their composition as to be undistinguishable from true mica-schists. On the whole, the two groups may be said to be composed essentially of felspar, quartz, mica, talc, hornblende, and chlorite—these ingredients having been originally deposited as silts and muds and sands, but subsequently altered in their internal structure so as to assume a crystalline or sub-crystalline character.

59. As a general rule, the gneiss group in the British Islands may be said to lie beneath the mica-schists, to be harder and less distinctly stratified, and to bear a closer resemblance to the granitic rocks with which it is intimately associated; hence the term *granitoid* or granite-like, by which gneissic compounds are frequently designated. The mica-schists, on the other hand, are less crystalline, more finely laminated, and present more frequent alternations of strata. In neither group are the strata very regular or continuous, and when broken up and contorted by intrusions of granite—which they frequently are—it is very difficult to trace either their lines of stratification or their order of superposition. The most prevalent rocks in these groups are—

Gneiss—an aggregate of quartz, felspar, and mica; occasionally containing garnets.

Syenitic Gneiss—of quartz, felspar, and hornblende.

Porphyritic Gneiss—gneiss either ordinary or syenitic, with large irregular macles of quartz or felspar.

Quartz-rock and Quartzite—granular aggregates of quartz, with occasional flakes of mica; the quartzites being indurated quartzose rocks of a more arenaceous aspect than quartz-rock proper.

Mica-schist—a fissile or laminated aggregate of mica and quartz, with occasional crystals of hornblende and garnet.

Talc-schist—of talc and quartz, and differs only in this respect from mica-schist.

Chlorite-schist—a greenish slaty rock of chlorite and quartz.
Hornblende-schist—a slaty rock, chiefly of hornblende, with felspar or quartz.
Actynolite-schist—a slaty foliated rock, chiefly of actynolite, with some admixture of felspar, quartz, or mica.
Primary Limestone—highly crystalline marbles, often containing veins and flakes of serpentine, chlorite, steatite, and the like.

The above are the principal rocks in the gneiss and mica-schist groups, and they often alternate and capriciously intermingle with each other. When they are massive and compact, they are spoken of as *rocks* (hornblende-rock); when they are capable of being split up or laminated, they are termed *schists* or *slates* (as mica-schist, hornblende-slate, &c.) The term *foliation* (*folium*, a leaf) is employed to express the irregular crumbled-like lamination that occurs among these rocks; and while we employ the term *schist* (Gr. *schisma*, a splitting or division) to embrace such strata as split up in this manner, the word *slate* ought to be restricted to those that exhibit regular cleavage, like roofing-slate.

60. The gneiss and mica-schist groups are widely distributed, being found flanking, less or more, all the principal mountain-chains in the world. They occur in the Highlands and Islands of Scotland, in the north of Ireland, in Brittany, along the flanks of the Pyrenees and Alps, in the Scandinavian and Oural chains, in the Himalaya and Altai ranges, in the mountains of Northern Africa, in the Andes and Brazilian sierras of South America, in the Cordilleras of Mexico, the Rocky Mountains, and along the entire length of the Alleghanies in North America. The physical aspect of these primary or metamorphic districts is bold, rugged, and barren. Thrown into lofty mountains by the granite, and often into abrupt and vertical positions, it is chiefly among gneiss and mica-schists that those deep glens and abrupt precipices occur, which give to highland scenery its well-known wild and picturesque effect. As already stated, the igneous rocks which upheave and disturb the gneiss and mica-schists are chiefly granitic; these upheaving, breaking through, and interstratifying with them in very complicated relations. Later igneous rocks—as porphyry, greenstone, and serpentine—are also found traversing these groups in the form of dykes and protruding masses; and occasionally still more recent effusions of trap are found passing through, not only the gneiss and mica-schists, but their associated dykes and veins of granite and porphyry.

61. The industrial or economic products derived from the gneiss and mica-schists are by no means numerous. Several of the metallic ores—such as tin and copper, and very rarely gold—

occur in veins traversing these strata. The limestones, from their highly saccharoid texture, and mottled and veined appearance, yield valuable marbles; and serpentine, when found in solid masses, produces also a very elegant material for internal decoration. Quartz-rock, when sufficiently pure, is quarried in several places for the potteries—the large blocks being used for grinding the calcined flints now so largely employed in the manufacture of the finer ware. Potstone, or the *lapis ollaris* of the ancients, of which jars and vases are sometimes manufactured, and amianthus or flexible asbestus, which may be woven into fabrics indestructible by fire, are both products of these rocks. One of the most valuable substances derived from the system is *graphite* or *plumbago*, so largely employed for writing-pencils, for polishing, for crucibles, and similar purposes. Garnets, rock-crystals, and other precious minerals, occur either in the strata themselves or in the veins that traverse them.

Clay-Slate Group.

62. Whatever obscurity may attach to the sedimentary origin of gneiss and mica-schist, there can be no doubt as to the true aqueous character of the clay-slates and their associated strata. The *Clay-slate Group*, so familiarly known by the bluish, greenish, and purplish roofing-slates of our towns, presents a vast thickness of fine-grained, fissile, argillaceous rock, of considerable hardness, and if not of a crystalline, at least of a glistening aspect. It seems to have been originally deposited as a fine clay or silt, and then to have undergone metamorphism in a less degree than the underlying mica-schists and gneiss rocks. The prevalent colours of slate are black, green, bluish, purplish, and mottled. Some varieties are hard and splintery, others soft and perishable. The texture, though generally close and fine-grained, is not unfrequently gritty and arenaceous, and passes into a sort of flaggy sandstone. The imbedded minerals are few; these being chiefly cubic iron-pyrites, chert or silicious concretions, crystals of hornblende, and chiastolite, a mineral occurring in long slender prisms, which cross and lie over each other in the mass of slate like the Greek letter χ (*chiastos*, crossed or marked with the letter χ, and *lithos*, a stone).

63. If gneiss and mica-schist were derived from the disintegration of older and granitic rocks, clay-slate seems to have been derived from the same source, and by a still further and finer comminution of the original rocky particles. In the clay-slates

the quartz and mica of the original rocks appear in minute grains and flakes, and the clay of the felspar appears as impalpable sediment, destitute of the potash and soda which entered into its crystallised condition in granite. All this bespeaks the long-continued action of atmosphere and water—atmosphere and water to waste and wear down, and rivers to transport the material to some tranquil sea of deposit. In course of time the soft sediment becomes consolidated; heat, pressure, or chemical agency subsequently changes its texture, and renders it hard and crystalline; and a still further alteration produces that peculiar structure in

Cleavage—a, Transverse; b, Coincident.

clay-slate known by the name of *cleavage*. What renders slate so peculiarly valuable, is its quality of being cleft or split into thin plates or layers; and this splitting takes place, not always parallel to the lines of stratification, as in the flagstones used for pavement, but in a direction (*a, b*) right through the beds, and often almost at right angles to them. This cleavage-structure is occasionally observable in other rocks of an argillaceous or clayey nature, but more especially in clay-slate; and its origin is still a matter of doubt among geologists. From its extreme regularity and resemblance to certain kinds of crystallisation, it is generally supposed to have arisen from chemical or magnetic forces acting in conjunction with pressure upon the clayey mass while in process of solidification—a supposition greatly strengthened by the fact that a similar structure has been produced in masses of clay by the artificial application of these forces. Indeed, pressure of itself is sufficient to induce in fine-grained homogeneous masses a structure very analogous to that of cleavage; hence the inclination of many geologists to ascribe cleavage, foliation, and analogous structures to this agency alone—acting at great depths and through long-continued periods.

64. Being intimately associated with the gneiss and mica-schist groups, the clay-slate partakes of all the upheavals and disruptions which have affected these strata. Though less crystalline in its texture, and not so much broken up by igneous intrusions, its beds are in many instances curiously bent and contorted, and generally rest at high angles on the flanks of our oldest mountain-chains. They are found along with mica-schist and gneiss in

almost all the regions already enumerated, and form valuable deposits in the Highlands of Scotland, in Cumberland, and in Wales. The scenery of clay-slate districts is often wild and picturesque; but their high elevation and cold clayey soils render them sterile and unproductive.

65. The industrial applications of clay-slate are numerous and well known. The hard fissile varieties have long yielded a most valuable roofing-material; the finer sorts are used for writing-slates and slate-pencils; and the thicker-bedded kinds are now largely employed as an ornamental stone for vases, tables, chimney-slabs, mosaic pavement, cisterns, and other architectural purposes. The clay-slate in many districts is traversed by metalliferous veins, and from these are obtained ores of tin, copper, lead, silver, and not unfrequently gold.

RECAPITULATION.

The system described in the preceding chapter consists of three principal groups—GNEISS, MICA-SCHIST, and CLAY-SLATE. The rocks composing these groups are all less or more indurated and crystalline, having their lines of stratification indistinct, and often altogether obliterated, and, as sedimentary strata, have evidently undergone some peculiar change in their internal structure. This change, or metamorphism, whether produced by heat, by pressure, or by chemical agency, has conferred upon them the term *Metamorphic rocks;* and by this designation they are now generally known to geologists, though there is no doubt that most of them are merely highly altered or metamorphosed strata of Silurian, Cambrian, and Laurentian age. As a system of formations, they are the deepest or lowest in the crust of the earth, and are therefore regarded as *Primary* or first-formed. They are known also as *Non-fossiliferous, Azoic,* or *Hypozoic* strata, from the fact that no distinct traces of plants or animals have yet been discovered in any part of the system. The terms metamorphic, primary, hypozoic, and non-fossiliferous, may be held as synonymous—the student ever bearing in mind that the nomenclature of Geology is at best but provisional or temporary, and must give way to new facts and the progress of discovery. As a general rule, the gneiss group lies beneath the mica-schist, and the mica-schist beneath the clay-slate; but there are frequent alternations of the rocks and schists composing the system, and these alternations are often rendered more complicated by the contortions and displacements produced by the intrusion of granitic outbursts. Though mineralogists have given to the rocks composing the system different names—as gneiss, syenitic-gneiss, horneblende-rock, horneblende-schist, mica-schist, talc-schist, stea-schist, chlorite-schist, &c.—it must be admitted that there are often a great similarity and frequent gradations among them. Whether this arises from their materials having been derived from the same set of older rocks, or, what is more likely, from the subsequent metamorphism which they have all alike undergone, it is difficult to say; but there is certainly a much closer family resemblance, so to speak, among the metamorphic strata than there is among the strata of any subsequent system. Though flanking and forming portion of most of the older mountain-chains, the primary strata do not occupy wide areas, but are tilted up at high angles, and compressed into a comparatively narrow space—producing rugged and abrupt scenery, less bold and bleak than granite, but wider and more irregular than that produced by later formations. In an economical point of view, the system is by no means unimportant. Slate and marble of various qualities are obtained from among the strata; and the ores of tin, copper, lead, silver, and gold, are extracted from the veins that traverse the system.

VII.

THE LAURENTIAN AND CAMBRIAN SYSTEMS, EMBRACING THE EARLIEST FOSSILIFEROUS SCHISTS, SLATES, AND ALTERED LIMESTONES.

66. As stated in the preceding chapter, the *Metamorphic System* is to be regarded merely as a provisional arrangement for such rocks as have undergone a high degree of metamorphism, and in which no fossils have as yet been detected. In districts where the mineral alteration has been less intense, fossils may naturally be looked for, and when such are found the containing strata are necessarily classed with the fossiliferous formations. In this way have arisen the *Laurentian* and *Cambrian* systems—the former so called from its vast development along the St Lawrence in Canada, and the latter from its forming the greater portion of Wales, the Cambria of our ancestors. These Laurentian and Cambrian strata, along with those now known as Silurian, constitute the *Greywacké* or *Transition* formation of the earlier geologists; the former term being a German word applied to certain slaty grits of a grey rusty colour which occur in the series, and the latter having reference to their fossiliferous character, and denoting (in their opinion) the transition of the world from an uninhabited to a habitable condition. The term Greywacké is now seldom employed, or employed only to designate a peculiar kind of rock; and the higher knowledge of modern science has all but exploded the idea of a transition period in the history of our planet. To Professor Sedgwick we owe the first definings of the Cambrian system, and to Sir William Logan, of the Canadian Geological Survey, we are indebted for those of the Laurentian; the two systems, in the present state of our knowledge, forming the lowest or earliest of the fossiliferous strata.

67. Before entering on the fossiliferous systems, however, it may be of use to the student to remind him that the department of his science having special reference to fossils is termed *Palæontology* (*palaios*, ancient; *onta*, beings; and *logos*, a discourse), or that

which treats of the ancient or former life of the globe; while the department more immediately concerned with the rocks and their physical relations is spoken of as *Lithology* (*lithos*, a stone; and *logos*). The palæontological and lithological aspects of a system are therefore two very different things, and convey much the same meaning as when we speak of the *stratigraphical* order of its rocks, and the *zoological* or *botanical* characters of its fossils. In describing the fossils of a system, the brief term *Flora* is usually employed to denote the general character of its plants, while the term *Fauna* is applied to that of its animal remains. To enter minutely into the subject of fossils would require on the part of the young student a more intimate acquaintance with the structure of living plants and animals than he is likely at this stage of his progress to possess. It may be useful, however, to remember that the plants are either *flowering* or *flowerless*—the former comprehending the true timber-trees, shrubs, pines, palms, and grasses, and the latter the lowlier club-mosses, ferns, equisetums, mosses, lichens, and sea-weeds; and that animals are either *vertebrate* (back-boned) or *invertebrate*—the former comprising the mammals, birds, reptiles, and fishes; and the latter the shellfish, crabs, insects, worms, starfish, sea-urchins, corals, sponges, and other lowly organisms. In a fossil state the harder and more durable portions are usually best preserved; hence the perfection of corals, shells, scales, spines, teeth, and bones, among animal structures, and trunks, branches, hard fruits, and dry leathery leaves among vegetables. Whatever the object preserved, there is either the real substance itself, the substance replaced by some mineral matter, a cast or mould of the substance, or it may be a mere mechanical impression, such as a foot-print or worm-burrow.

68. Beginning with the LAURENTIAN as the oldest fossiliferous system, it consists in the typical district of the St Lawrence and Laurentide Mountains of a vast thickness (30,000 feet or thereby) of crystalline strata like the gneiss, quartz-rocks, marbles, and serpentines of the Scottish Highlands, or more perhaps like the still harder and more granite-looking schists of the Scandinavian Mountains. There are no sandstones, nor shales, nor limestones, in the proper sense of the term. All these have been converted by heat, pressure, and chemical action into sparkling crystalline rocks; lines and layers of stratification are obscure and often altogether obliterated; veins and eruptive masses are frequent; and altogether the whole system wears the aspect of a vast and venerable antiquity. It is sometimes divided into

UPPER LAURENTIAN or LABRADOR SERIES, and
LOWER LAURENTIAN;

but the whole is so metamorphosed that subdivision is usually impossible, and for all ordinary purposes it may be considered as a vast alternation of granitoid schists, gneisses, quartzites, limestones, and serpentines.

69. In rocks that have undergone so much change of texture and structure, fossil remains are scarcely to be expected—the metamorphism that induced the crystalline character being sufficient to obliterate, or all but obliterate, every trace of organic existence. And yet in some serpentinous limestones diligent research, aided by the microscope, has recently (1863) detected the presence of organic structure—this structure being many-celled calcareous masses elaborated by *foraminifera*, or the lowest forms of animal life. This cell-structure, visible only through the microscope, is so distinctive as compared with any mere mineral texture that no doubt remains among competent observers as to its animal origin; and thus these old Laurentian schists have been erected into an independent fossiliferous system—the lowest or earliest with which geology is yet acquainted. The organism, *Eozoön Canadense*, or Dawn-animalcule of Canada, occurs in coral-like masses, forming layers or bands in the thick-bedded serpentines of

Eozoon Canadense.—1, The layers, natural size; 2, The Tubuli, magnified 100 diameters.—CARPENTER.

the St Lawrence, and has also been more recently detected in rocks apparently of the same age in Ireland and Bohemia. That similar organisms may be discovered in the serpentines of other regions is not improbable; but, in the mean time, where fossils do not occur and where the stratigraphical sequence from Silurian into Cambrian and from Cambrian into a still older series is not very obvious, it will be better to regard such unresolved schists merely as *Metamorphic*, without attempting to define them either as Cambrian or Laurentian.

70. Above the Laurentian, and having, on the whole, undergone less metamorphic change, occur the slates, schists, grits, and crystalline limestones of the CAMBRIAN system, typically displayed in the mountains of Western Wales—the "Cambria" of our ancestors. In consequence of this less degree of metamorphism, not only is stratification more distinct, but the sedimentary texture is less altered and the contained fossils more readily procured and legible. Like other formations, the Cambrian may vary in composition in different regions, sometimes being more slaty, and at others more schistose and crystalline, but, on the whole, slates, schists, grits, and altered limestones, to the thickness (as in Wales) of 15,000 or 20,000 feet, may be taken as the usual composition of the system. In the typical district of Wales it is usual to arrange the strata into a lower and upper zone—the upper insensibly merging into the Silurians above—thus:—

UPPER. { Tremadoc slates; dark earthy slates with pisolitic ore.
         { Lingula flags; micaceous flagstones and shales.

LOWER. { Harlech grits; sandstones and silicious grits.
        { Llanberis slates; slates with intercalated grits.

As in all the older systems, the Cambrian is frequently traversed by veins and eruptive masses; and hence its importance as a repository of the useful minerals and metals.

71. As regards organic remains, the system has yielded impressions of sea-weeds (*fucoids*), undetermined zoophytes (*Oldhamia*), brachiopodous mollusca (*lingula*), chambered mollusca (*orthoceras, cyrtoceras*), crustacea (*paradoxides, olenus, agnostus*, and other

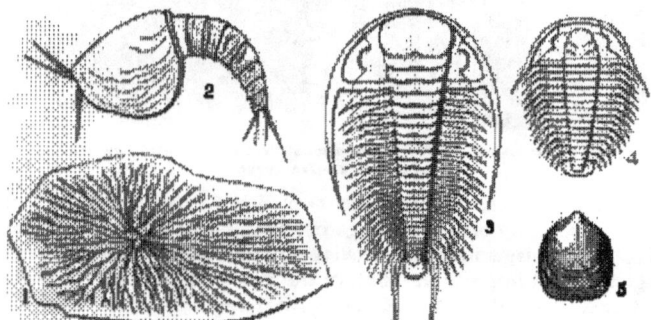

1, Oldhamia Radiata. 2. Hymenocaris. 3. Paradoxides; 4. Olenus; 5. Lingula.

trilobites), tracks and burrows of annelids (*arenicolites*), and other tracks and impressions less distinct and ascertainable. On the whole, the flora and fauna are of lowly nature, and but sparingly scattered through the strata; and though usually exhibiting the same genera as the lower Silurians, are yet found, when critically examined, to belong to different species. Respecting the extent and distribution of the Cambrian system, our knowledge is yet very partial and imperfect, but strata containing similar fossils with those of Wales have been found in Bohemia, Scandinavia, Canada, the United States of North America, and along the lower flanks of the Andes.

72. The physical aspects of Laurentian and Cambrian regions are much the same as those that characterise the Metamorphic system—bold, rugged hill-ranges, abounding in steep precipices and splintery peaks; deep narrow gorges; and all that irregularity of surface that belongs to slaty and schistose formations when weathered and worn down by meteoric agency. The Northern Hebrides, Scandinavia, the Laurentian mountains, our own Wales, and the mountains of Cumberland, may be taken as types of the physical geography peculiar to Laurentian and Cambrian districts.

73. As industrial repositories both formations yield slates of unrivalled quality, serpentines, ores of iron, tin, copper, silver, and other metals. Indeed, it is chiefly in these earlier formations, whether called Metamorphic, Laurentian, Cambrian, or Silurian, that the richest metalliferous veins occur, nature having had longest time, as it were, to elaborate these ores from the solutions that are incessantly percolating the chinks and fissures of the crust.

RECAPITULATION.

In the preceding paragraphs a brief outline is given of the Laurentian and Cambrian systems—the two lowest fossiliferous formations yet discovered in the crust of the earth. For the most part highly metamorphosed, and consisting of crystalline slates, schists, grits, and serpentines, they were till recently regarded as portions of the Metamorphic Rocks, and have only been separated and erected into independent systems through the discovery of the fossils above alluded to. In this way all the metamorphic rocks may be regarded as "unresolved schists," awaiting the detection of fossils in their less altered portions, to enable geologists to assign them either to the Cambrian, to the Laurentian, or to some still deeper-seated and earlier life-system. In the mean time there can be no error in classifying all crystalline slates and schists which have yielded no fossils as *Metamorphic*, and this method will less mislead than any attempt to rank them, in the absence of fossil evidence, either as Silurian, Cambrian, or Laurentian. The *Laurentians* proper are largely developed in the Laurentian mountains of Canada, and equivalent rocks are thought to occur in the further Hebrides, in Western Ireland, in Bohemia, and Scandinavia; while the *Cambrians* and strata of the same age have been examined in Wales, Cumberland, north-west of Scotland, Ireland, Bohemia, and North America. Of vast antiquity, and having undergone great changes through pressure, heat, and chemical agency, they are generally rich in metalliferous veins; and it is from rocks of this age that a large proportion of the ores of iron, copper, tin, silver, and other valuable metals are obtained. And it is also owing to this antiquity, to their slaty and schistose structure, and to the long ages during which they have been subjected to meteoric and aqueous waste, that Laurentian and Cambrian rocks confer on the regions in which they occur their wild, rugged, and picturesque scenery.

· VIII.

THE SILURIAN SYSTEM, EMBRACING THE LOWER AND UPPER
SILURIAN GROUPS, OR THE LLANDEILO, WENLOCK,
AND LUDLOW SERIES.

74. IN whatever condition the metamorphic rocks were at first laid down in the seas of deposit, we have seen that a common crystalline aspect now pervades the whole series, and that the usual alternations of sedimentary matter are all but obliterated. We cannot say, for example, which stratum was originally of clay-silt, which of sand, or which of gravel. All these distinctions are effaced, and we cannot arrive at any satisfactory conclusion as to the waves and tides and currents by which they were aggregated, or the nature of the seas in which they were deposited. The case, as has been shown, is somewhat different in the Laurentian and Cambrian systems, where deposition and fossils are not altogether obliterated; but it is widely different in the Silurian, in which every stratum and alternation of strata are, for the most part, distinctly traceable. Beds of slaty sandstone, pebbly conglomerate, shaly mudstone, and limestone, follow one another in frequent succession, and present so slight a change in their mineral structure that we can readily judge of the conditions under which they were deposited. Some of the sandstones are finely laminated, and bear evidence of tranquil sediment; others are ripple-marked, and testify to the presence of tides or gentle currents; some are marked by the trails or pierced by the burrows of sand-worms; while others are pebbly conglomerates, and bespeak the existence of waves and gravel-beaches, such as we witness at the present day. Of the shales or argillaceous beds, some have evidently been thrown down in deep water as soft black mud, while others have been formed in shallower bays, and contain a certain admixture of sand, with sea-shells, such as are found at no great depth from the shore. Of the limestones or calcareous strata, many are replete with the remains of corals and shells, and recall the existence of

seas in which the coral-polype reared its reefs, and shell-fish congregated in beds like the oyster and mussel of our own times. Indeed, the abundant presence of fossil zoophytes, corals, molluscs, and crustaceans, tells of varying conditions of water and sea-bottom, of light and heat, of tribes that secreted their nutriment from the ocean, or preyed on each other; and generally of a state of things different from, it may be, but still analogous to, that which we perceive in existing nature.

Lower and Upper Silurian Groups.

75. The system which contains evidence of these varied conditions, consists essentially of argillaceous, arenaceous, and calcareous strata. Dark-coloured laminated shales, shales with concretions of limestone, beds of calcareous flagstone, thick-bedded sandstones and pebbly conglomerates, finely laminated micaceous sandstones and shales, and impure clayey limestones, and limestones of a concretionary structure, may be said to constitute the entire system. These strata, or rather system of strata, being very fully developed in that district of country between England and Wales, anciently inhabited by the Silures, the term *Silurian* has been applied to them, and their succession very carefully worked out, both in its mineral and fossil aspects. In this investigation, the system has been arranged into an upper and lower group, or, taking the typical district—the Siluria of the ancient Britons—into the *Llandeilo*, *Wenlock*, and *Ludlow* series, as represented in the subjoined synopsis:—

UPPER SILURIAN.

LUDLOW SERIES.
- Finely laminated reddish and greenish sandstones and shales, locally known as "Tilestones" (in part, base of the Devonian System).
- Micaceous grey sandstone, in beds of varying thickness.
- Argillaceous limestone (Aymestry limestone).
- Shale, with concretions of limestone (Lower Ludlow).

WENLOCK SERIES.
- Concretionary limestone (Wenlock limestone).
- Argillaceous shale in thick beds.
- Shelly limestone and sandstone (Woolhope and Mayhill).
- Gritty sandstones and shales (Upper Llandovery).

LOWER SILURIAN.

LLANDEILO SERIES.
- Grits and sandy shales (Lower Llandovery).
- Thick-bedded white freestone (Caradoc sandstone).
- Dark calcareous flags and slates (Bala beds).
- Dark-coloured calcareous flags, bands of limestone, and gritty flagstones (Longmynd or "Bottom rocks").

The preceding synopsis represents a thickness of about 8000 feet, and contains many alternations and gradations from freestone to sandy flags, from flagstones to shales, and from shales to calcareous flags and limestones of varying thickness and purity.

76. The fossils of the Silurian system are eminently marine, and consist of numerous species and genera of zoophytes, radiata, molluscs, annelids, and crustacea. Traces of fishes are found on the uppermost verge of the system, or in beds, which by some are considered as the basis of the Old Red Sandstone; and fuci, or sea-weeds, as well as the seed-spores and stems of plants apparently allied to the lycopodium or club-moss, have also been detected in the same strata; but we have as yet no evidence of any terrestrial fauna. Of course, it is not to be supposed that every portion of the system has been fully investigated: the strata as yet examined may have been deposited in deep water; and not till those deposited along the shores and in the estuaries of the rivers which carried down the sand and mud of the period have been equally well explored, can we pronounce with certainty either as to the kind or amount of fossil remains. As it is, numerous genera of a varied and prolific sea-fauna have been detected, and these are invested with a high interest, not only from their great antiquity, but from the beautiful state of preservation in which they frequently occur. And here let the student impress on his mind the fact, that, though among the earliest forms of life, there is in their structure no imperfection or trial-work. The corals of the Silurian seas, the shell-fish and crustacea of this primeval period, are as complex in their organisation,

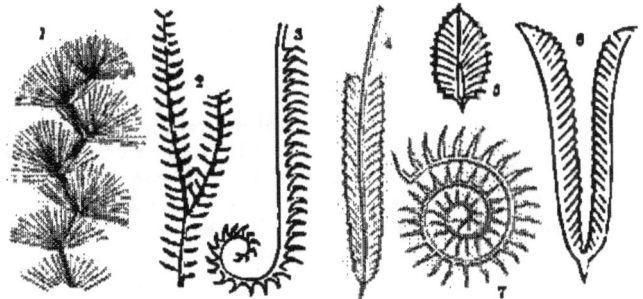

1. Oldhamia antiqua; 2. Protovirgularia; 3. Graptolites; 4, 5. Diplograpsus; 6. Didymograpsus; 7. Rastrites.

and as perfectly fitted for the functions they had to perform, as the corals and shell-fish and crustacea that now throng

ITS FOSSILS.

existing waters. Among the more characteristic fossils of the period may be noticed the *graptolites* (*grapho*, I write, and *lithos*, a stone), a peculiar family of zoophytes, so called from their resemblance to the sea-pens (*sertularia* and *virgularia*) of our own seas. These zoophytes seemed to have thronged the muddy bottom of the Silurian waters, and are highly distinctive of the lower portion of the system. Among the corals and coralloid remains of the period, there are also many peculiar genera, remarkable either for their sponge-like appearance, or for the cup-like form of their structure. From the form or arrangement of their pores, these corals are known by such names as *cyathophyllum*, or cup-coral; *astræa*, or star-coral; *heliolites*, or sun-coral; *favosites*, or honeycomb coral; and *catenipora*, or chain-

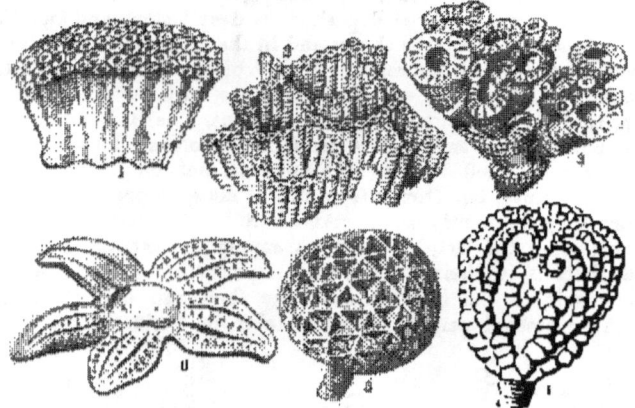

1, Heliolites; 2, Catenipora; 3, Cyathophyllum; 4, Tascrinus; 5, Cystidea; 6, Palæaster.

pore coral. Among the radiate or rayed animals, whose structure resembles that of the star-fish, several well-marked groups have been found in Silurian strata. Among the most characteristic of these may be noticed the *encrinites*, or lily-like radiata (Gr. *krinon*, a lily), whose calcareous skeletons often constitute the main mass of certain limestones. True star-fishes, allied to the *uraster* and *comatula* of our own seas, have also been discovered; and certain remarkable bladder-shaped forms, called *cystideæ* (*kystos*, a bladder), which seem to approach to the sea-urchins in structure. As an encrinite, with its numerous arms and feathery branches, may be regarded as a star-fish fixed to the bottom by a jointed and flexible stalk, so may a cystidean, with its spherical body, composed of numerous plates, be considered a sea-urchin attached to the bottom by a similarly-jointed column. Among the molluscs or shell-fish found in Silurian strata, there are the representatives of many existing orders—bivalves allied to the cockle and pecten, others to the mussel; whorled univalves like the periwinkle, spirals like the pelican's foot and turritella; chambered shells coiled up like the pearly nautilus, and others,

1. Lingula. 2. Rhynconella; 3. Pentamerus; 4. Euophomena; 5. Spirifera; 6. Murchisonia; 7. Orthoceras; 8. Lituites; 9. Maclurea.

massive and straight, to which we have no existing analogue. Of the bivalves, *terebratula, spirifer, orthis,* and *lingula,* are the

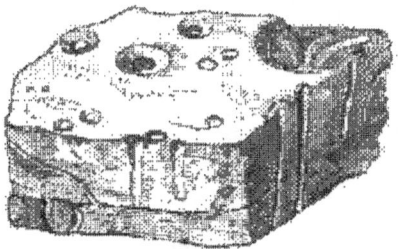

Worm-burrows (Scolithus linearis) in Sandstone.—MURCHISON.

most characteristic; of the univalves, *pleurotomaria, Murchisonia,* and *euomphalus;* and of the chambered shells, *lituites, orthoceras,* and *phragmoceras*. Of annelids—that is, ringed or worm-like creatures, so called from *annulus*, a ring,—there are the tracks and burrows of sea-worms in the sandstones and sandy shales, known as *arenicolites* and *scolites;* the calcareous casts or shell-like cases of spirorbis and serpulæ—the latter being known by the term *serpulites*, from their resemblance to the *serpula* of existing seas. By far the most curious and abundant, as well as the most charac-

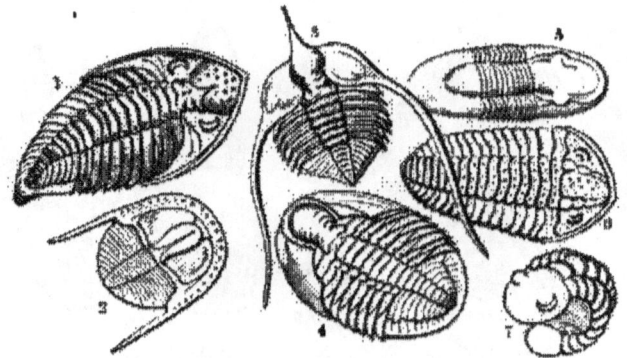

1. Phacops; 2. Trinucleus; 3. Ampyx; 4. Olygia; 5. Illænus; 6. Calymene; 7. Calymene rolled up.

teristic of Silurian fossils, are the crustaceans, termed *trilobites*, from the three-lobed-like figure of their bodies. These creatures seem to have swarmed in the Silurian waters in numerous genera and species, just as shrimps and prawns swarm in the shallow

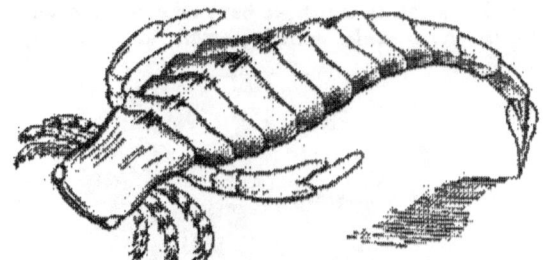

Dorsal aspect of Slimonia acuminata. From the Upper Silurian or Passage Beds of Lanarkshire.

seas of our own day. Different genera and species are distinguished by such terms as *asaphus* (obscure), *calymene* (concealed), *ilænus*, *ogygia, phacops, trinucleus,* &c., and may be fairly considered as one of the leading orders of life that appeared during the silurian epoch. Besides these crustaceans we have the higher and larger forms of *eurypterus* (broad-paddle), *stylonurus* (style-tail), *pterygotus* (paddle-ear), *Slimonia* (after its discoverer, Dr Slimon) —first appearing in the uppermost beds of the system, and passing upwards in greater numbers and variety into the succeeding system of the Old Red Sandstone. As already stated, remains of fishes are found in the uppermost beds of the system, but these are regarded by Sir Roderick Murchison as marking the dawn of the Devonian rather than the close of the Silurian era—"a long early period, in which no vertebrated animals had been called into existence." This opinion must be received, however, as indicating the paucity of such remains rather than their total absence: and for the final grouping of the "Tilestone" beds either as Silurian or Devonian, we must wait more extended research and the progress of discovery.

77. Respecting the extent of country occupied by silurian strata, we have as yet no very accurate information. As before mentioned, they are most typically displayed in the district of country between England and Wales ; the lower series occurs in a broad band along the entire south of Scotland, from St Abb's Head to Portpatrick, and appears also in Cumberland, Westmoreland, and along the south-east coast of Ireland. The system is found in Scandinavia, in Russia and the Ourals, and very characteristically in Silesia and Bohemia. Silurian strata have also been investigated in the south of France, in Asia Minor, in Australia and New Zealand, and in North and South America, and, as the progress of research advances, will no doubt be discovered in other regions. In all these districts the system is marked by the same, or nearly the same, genera of fossils, and though the strata may differ very greatly in a mineralogical point of view —shales, for example, passing from soft disintegrating mudstones to hard fissile slates, sandstones passing from laminated sandstones to jaspery conglomerates, and limestones from calcareous marls to concretionary cornstones—still the moment a geologist detects graptolites, trilobites, lingulæ, orthidæ, and the like, he can have no doubt as to his position among true silurian strata.

78. The igneous rocks associated with the system are partly imbedded or contemporaneous, and partly eruptive. The imbedded traps are chiefly felspathic ash and tufa of a mixed mineral character, and have evidently been laid down in these primeval seas, sometimes in the state of overspreading or molten lava, and sometimes in the state of showers of scoriæ and ashes. The

eruptive rocks are principally felspathic—felspathic greenstones, felspar rock, and felspar porphyry. In many instances, as in Wales and the south of Scotland, they have rendered the strata partially metamorphic, converting shales into good useful roofing-slates, sandstones into quartzite, and clays into hard jaspery hornstone. The upheavals and contortions resulting from their eruptions produce, on the whole, a varied and picturesque scenery, less abrupt and bold than that of primitive districts, and yet more diversified by hill and dale, by ravine and river-glen, than that of later or secondary periods.

79. In an industrial point of view, the rocks of the silurian system are of no great importance. Roofing-slate of various quality is obtained from the series, but of inferior value to that of the true clay-slate; flagstones are quarried in some districts, though inferior to those of the old red sandstone; freestone for building purposes is also a local product; and limestone for mortar and manure is quarried and burnt in most silurian countries. The veins that traverse the system are in general metalliferous, and from these, ores of mercury, copper, lead, silver, and gold are extracted. Indeed, according to Sir Roderick Murchison, "the most usual original position of gold is in quartzose veinstones that traverse altered palæozoic slates, frequently near their junction with eruptive rocks. Sometimes, however, it is also shown to be diffused through the body of such rocks, whether of igneous or of aqueous origin. The stratified rocks of the highest antiquity, such as the oldest gneiss or quartz rocks, have very seldom borne gold; but the sedimentary accumulations which followed, or the Silurian, Devonian, and Carboniferous (particularly the first of these three), having been the deposits which, in the tracts where they have undergone a metamorphosis or change of structure by the influence of igneous agency or other causes, have been the *chief* sources whence gold has been derived."

RECAPITULATION.

In the preceding chapter we have presented an outline of the Silurian System, chiefly as known to British geologists. Originally designated the *Greywacké* or *Transition* formation, and but imperfectly defined and little understood, these strata have now undergone a most minute and careful survey as regards both their palæontology and their order of superposition. They are largely developed in various countries, both in the Old and in the New World, and typically so in the district between England and Wales anciently inhabited by the Silures; hence the designation by Sir R. Murchison, their first and most ardent investigator, of the SILURIAN SYSTEM. The system, though consisting, in the main, of alternations of flagstones and sandstones, of argillaceous and calcareous shales, of clayey limestones, and limestones of a concretionary structure, has been divided into *lower* and *upper* groups, and these groups again, in the typical district, into the *Llandeilo*, *Wenlock*, and *Ludlow* series. In the several series, abundant traces of lowly life have been detected, and numerous species of zoophytes, radiata, mollusca, annelida, and crustacea, figured and described. Remains of fishes have also been found in the upper beds, but these are regarded as marking the dawn of the Old Red Sandstone epoch, rather than as belonging to the close of the Silurian. Adhering to this view as a mere provisional line of distinction, we obtain a well-marked palæontological basis for the Old Red Sandstone;—and can view the graptolites; the favosites and heliolites; the actinocrinites, the marsupites, and cystidæ; the lingulæ, terebratulæ, and orthidæ; the lituites and orthoceratites; the serpulites and tentaculites; the asaphus, calymene, trinucleus, and other trilobites, as the peculiar and distinctive fauna of the silurian era. These creatures are all of true marine habitat; and, coupling this with the facts of ripple-mark, and with frequent alternations of shales, which were originally sea-silt—of sandstones, which point to sandy shores—of conglomerates, which speak of gravel-beaches—and of limestones, that tell of shell-beds and coral-reefs,—we are carried back through the lapse of ages to a series of seas and bays and estuaries, in which the operations of life and development went forward, deepening and spreading and multiplying even as they do now. The silurian strata seem to be extensively developed in most countries of the world, and, though affording no important mineral product of themselves, are, along with the other palæozoic rocks, thought to be the chief repositories of those auriferous veinstones whose golden metal has ever been the coveted treasure of man, whether savage or civilised.

IX.

THE OLD RED SANDSTONE OR DEVONIAN SYSTEM, EMBRACING THE GREY FLAGSTONE, THE RED CONGLOMERATE, AND THE YELLOW SANDSTONE GROUPS.

80. TAKING the Coal-Measures as a sort of middle point, there is generally found in the British Islands one set of reddish sandstones lying beneath, and another set lying immediately above them. By the earlier geologists, the lower set was designated the *Old Red Sandstone*, and the upper the *New Red Sandstone;* and though the progress of the science has rendered it necessary to impose certain limitations on these terms, they are still sufficiently distinctive and easily remembered. The Old Red Sandstone system, therefore, may be held as embracing the whole series of strata which lies between the Silurian on the one hand and the Carboniferous on the other. Certain portions of the system are peculiarly developed in Devonshire,—hence the term *Devonian*, which is generally employed as synonymous with the earlier and more descriptive one of "Old Red Sandstone."

81. The Old Red Sandstone, as the name sufficiently indicates, consists of a succession of sandstones, alternating with subordinate layers of sandy shale and beds of concretionary limestone. The sandstones pass in fineness from close-grained fissile flagstone to thick beds of coarse conglomerate, and the shales from sandy laminated clay to soft flaky sandstone. The whole system is less or more coloured by the peroxide of iron—the shades varying from a dull rusty grey to a bright red, and from red to a fawn or cream-coloured yellow. Many of the shades are curiously mottled—green, purple, and yellow—and present an aspect which, once seen in the field, is not soon forgotten. On the whole, shades of reddish colour may be said to pervade the system, unless in some of the lower slaty bands, which present a dark and semi-bituminous aspect. The slaty bands of limestone are locally known as *flagstones* and *tilestones;* the conglomerates, which are

merely solidified gravel and shingle, are fancifully termed *puddingstones*—the pebbles being mingled through the mass like the fruit in a plum-pudding; and the limestones, from their silicious or concretionary texture, are generally known by the name of *cornstones*. The shales are occasionally soft and friable, and in this state are by some termed *marls*, but from their containing no lime the name is by no means appropriate.

82. Starting from the flaggy beds which top the silurian rocks, and contain the spines, scales, teeth, and other remains of fishes, the following may be taken as the order of succession among the Old Red Sandstone strata, at least as typically developed in the British Islands:—

YELLOW or UPPER GROUP.
1. Yellow sandstones, generally fine-grained, but including detached pebbles, and alternating with layers of mottled shales. *Aquatic and terrestrial plants; few zoophytes or shell-fish, but numerous fishes.*
2. Marine limestones and sandy schists, like those of Devonshire. *Abundant remains of corals, shells, and crustacea, but few fishes.*

RED or MIDDLE GROUP.
3. Coarse red conglomerates interrupted by occasional beds of chocolate-coloured quartzose sandstone. *Occasional fish-remains and plants.*
4. Red sandstones, generally in thick beds of a dull brick red, enclosing detached pebbles of quartz and other rocks. Conglomerate beds, layers of greenish shale, and beds of concretionary limestone, are occasionally interstratified. *Organic remains rather rare, and not very distinct.*

GREY or LOWER GROUP.
5. Grey or reddish rusty grey sandstones with enclosed pebbles; and thick masses of conglomerate interstratified with greyish flagstones and sandy shales. *Organic remains rare, and imperfectly preserved.*
6. Dark grey micaceous flagstones and tilestones, with occasional flaggy schists of a dark bituminous aspect. *Numerous impressions of aquatic and land plants; shell-fish doubtful, but abundant remains of crustacea and fishes.*

The preceding synopsis represents the usual order of the system as it occurs in Scotland and England, though no one district presents an entire suite from the lowest to the highest strata.

83. The organic remains of the Old Red Sandstone, though not profusely scattered through the system, are of high and increasing interest, inasmuch as they furnish distinct evidence of terrestrial vegetation, as well as the earliest traces of vertebrate life on our globe. Among the flagstones and sandstones, and also

ITS FOSSILS. 87

among some of the more laminated shales, we have impressions of *fuci*, or sea-weeds (*chondrites* and *zosterites*); of marsh-plants,

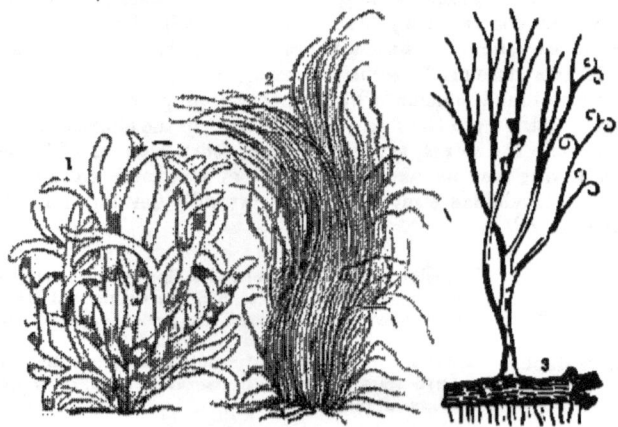

1. Fucoid (Roxburghshire); 2. Zosterites (Forfarshire); 3. Psilophyton (Canada).

apparently allied to the equisetum, the bulrush, and sedge (*jun-*

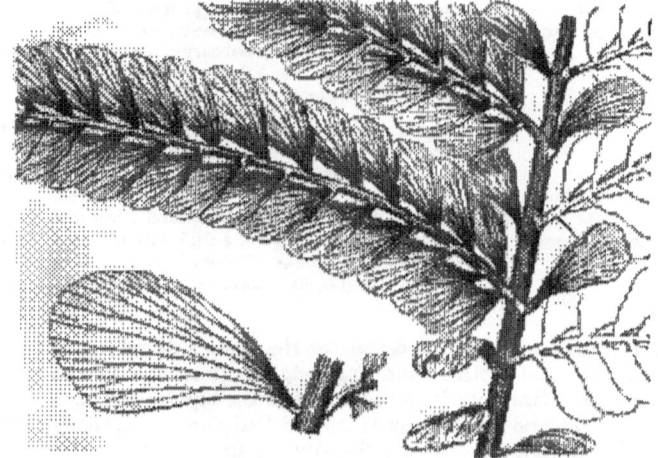

Adiantites Hibernicus.—Yellow Sandstone Series of Ireland.

cites); and of land-plants akin to the tree-ferns (*adiantites*), calamites, and lepidendron. On the whole, within the area of Great

Britain and Russia (where the system is largely developed) the vegetable remains occur in a fragmentary and carbonised state, as if they had been drifted from a distance to the sea of deposit; but in Canada the plants are much better preserved, and so abundantly as to form, in some instances, thin seams of coal. Among the flaggy shales of the lower groups, as in Caithness, there occur dark bituminous bands of considerable importance; but these, it would appear, are the products of animal rather than of vegetable decay. As a whole, the system seems by no means fertile in plant-remains; and even of such as do occur (with the exception of some fucoids and ferns), the botanist has yet been unable to render any certain interpretation. The Fauna of the system, on the other hand, is much more abundant and peculiar. Taking the strata as more especially investigated in Scotland, England, Ireland, and Belgium, it may be safely asserted that fishes of remarkable structure and organisation constitute the characteristic fossils. It is true that we have many genera of corals, *cyathophyllum, arachnophyllum, favosites*, &c.; of encrinites, *actinocrinus, cyathocrinus;* of shells, *spirifera, megalodon, calceola, stringocephalus, clymenia*, &c.; of peculiar crustaceans, like the *pterygotus (pteryx*, a wing, and *ous, otos*, an ear); but that profusion of corals, shells, and trilobites,

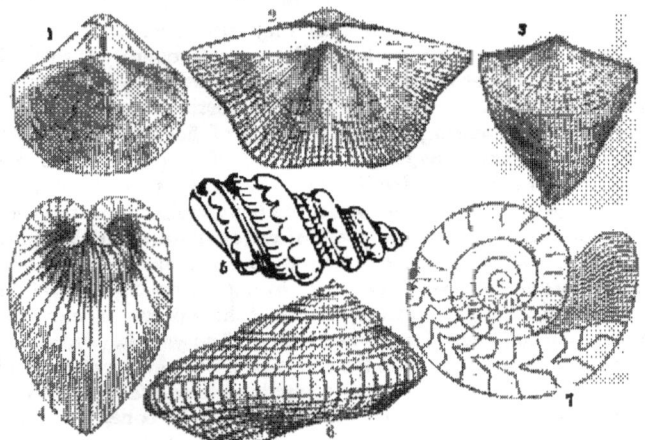

1. Stringocephalus; 2. Spirifera; 3. Calceola; 4. Megalodon; 5. Murchisonia; 6. Pleurotomaria; 7. Clymenia.

which thronged the silurian seas, is here superseded by other orders and other arrangements. One of the most notable of these new arrangements is the prevalence of crustacean forms (*pterygotus, stylonurus, eurypterus*, &c.), some of them, like the pterygoti of Forfarshire, measuring from four to six feet in length, and otherwise gigantic in proportion. The fishes of the

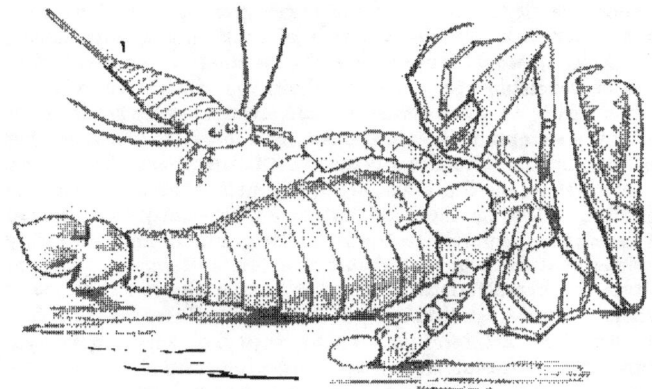

1. Stylonurus Powrei ; 2. Pterygotus Anglicus (ventral aspect). From the Lower Old Red of Forfarshire.

period are peculiar, inasmuch as they are covered with bony plates, or with hard enamelled scales ; are frequently furnished with bony spines or external defences ; and are many of them of forms widely different from the fishes of existing seas.

84. Without entering into the anatomy of fishes, we may here explain a few terms frequently made use of by geologists in describing those remains. Fossil, like living, fishes, are either *osseous* or *cartilaginous* ; that is, have either a bony skeleton like the haddock and salmon, or one composed of mere cartilage like the shark and skate. As the scales or external covering are the best preserved portions of fossil fishes, they have been arranged into four great orders, according to the structure of these parts—namely, the ganoid, placoid, ctenoid, and cycloid. 1. The *ganoid* (Gr. *ganos*, splendour) are so called from the shining surface of their enamelled scales. These scales are generally angular, are *regularly* arranged, entirely cover the body, are composed internally of bone, and coated with enamel. Nearly all the species referable to this division are extinct ; the sturgeon and bony-pike of the North American lakes are living examples. 2. The *placoid* (*plax*, a plate) have their skins covered irregularly with plates of enamel, often of considerable dimensions, but sometimes reduced to mere points, like the

F

shagreen on the skin of the shark, or the prickly tubercles of the ray. This order comprises all the cartilaginous fishes, with the exception of the sturgeon. 3. The *ctenoid* (*kteis, ktenos,* a comb) have their scales of a horny or bony substance without enamel, and jagged on the posterior edge like the teeth of a comb. The perch may be taken as a living example of this division. 4. The

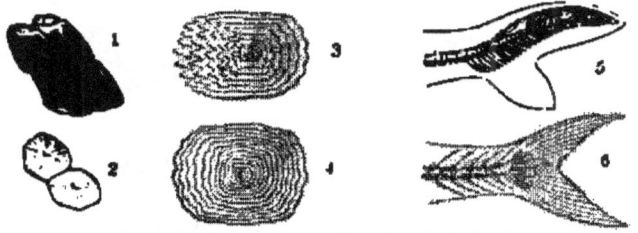

1. Ganoid ; 2. Placoid ; 3. Ctenoid ; and 4. Cycloid Scales ;
5. Heterocercal ; 6. Homocercal Tail.

cycloid (*kyklos,* a circle) have smooth, bony, or horny scales, also without enamel, but entire or rounded at their margins. The herring and salmon are living examples of this order, which embraces the majority of existing species. Besides these distinctions, it is also usual to recognise fossil fishes as heterocercal and homocercal ; that is, according as their tails are unequally or equally lobed. Thus, in *heterocercal* species (*heteros,* different, and *kerkos,* a tail) the tail is chiefly on one side, like that of the shark and sturgeon, the backbone being prolonged into the upper lobe ; in *homocercal* species (*homos,* alike) the lobes of the tail are equal or similar, as in the salmon and herring. In palæontology this distinction, as will afterwards be seen, is an important one, all the fishes of the palæozoic periods being heterocercal, the equally-lobed and single rounded tails being characteristics of more recent and existing species.

Cephalaspis Lyellii—Side and Dorsal Aspects.

85. Of the more characteristic FISHES, we may notice the *cephalaspis*, or buckler-head (*kephale*, the head, and *aspis*, a buckler), so named from the shield-like shape of its head; the *coccosteus*, or berry-bone (*kokkos*, a berry, and *osteon*, a bone), so called from

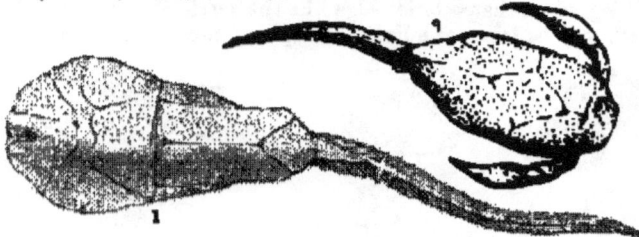

1. Coccosteus; 2, Pterichthys.

the berry-like tubercles which stud its bony plates; the *pterichthys*, or wing-fish (*pteryx*, a wing, and *ichthys*, a fish), which receives its name from the peculiar wing-like appendages

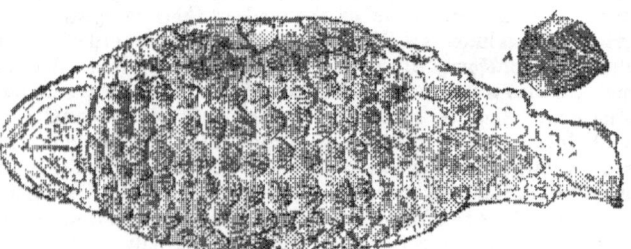

Holoptychius. A. Detached Scale.

attached to its body; the *holoptychius*, or all-wrinkle (*holos*, entire, and *ptyché*, a wrinkle), so termed from the wrinkled surface of its large enamelled scales; *osteolepis*, or bone-scale (*osteon*,

Diplacanthus gracilis.—Forfarshire.

a bone, and *lepis*, a scale); *dipterus*, or double-fin; *diplacanthus*, or double-spine; *phaneropleuron*, or obvious-rib, and so forth—

all of them receiving their names from some marked and peculiar external feature. These fishes seem to have thronged the waters of the period, and their remains are often found in masses, as if they had been suddenly entombed in living shoals by the sediment which now contains them. Occasionally, only detached scales are found, as if these had been drifted about on the shores of deposit,

Phaneropleuron.   A. Detached Scale.—Dura Den.

and at other times a spine is all that bears evidence of their existence. Any fragment of a fossil fish, whether bone or scale, is termed an *ichthyolite* (*ichthys*, a fish, and *lithos*, a stone); and the detached fin-spines are known by the name of *ichthyodorulites* (*ichthys*, a fish; *doru*, a spear; and *lithos*, a stone). Of the existence of REPTILES during the Old Red Sandstone period, we have as yet no unquestioned evidence. It is true that footprints, scutes, and bones occur in the sandstones of Lossiemouth, in Elginshire, which were, till recently, regarded as Upper Devonian; but now, partly on lithological, and partly on palæontological grounds, this opinion has been controverted by some of our most competent authorities, and these sandstones and their fossils (*telerpeton, staganolepis*, &c.) have, in accordance with this view, been removed to the Triassic era.

86. The igneous rocks more intimately associated with the system are greenstone, clinkstone, compact felstone, felstone-porphyry, claystone-porphyry, amygdaloid, and other varieties of felspathic trap. Unless in the lower group, these traps are rarely interstratified with the sandstones, and in this respect present a striking difference from the tufas and ashes which often alternate with the strata of the Silurian and Lower Carboniferous systems.

They occur chiefly as upheaving and disrupting masses, and are themselves frequently cut through by later dykes of greenstone, felspar, and porphyry—thus seemingly indicating a cessation of volcanic action during the main deposition of the Old Red Sandstone, but a period of great activity and disturbance both at its commencement and at its close. Granitic outbursts are rare in connection with the Old Red; and it may be assumed, as a general rule, that the period of the granite had given way to that of the trap, with its multifarious compounds. The physical features of red sandstone districts in Britain are generally diversified and irregular—the hills being less bold and precipitous than those of primitive districts, and more lofty and irregular than those of the later secondaries. Where the strata are unbroken by trap eruptions, the scenery is rather flat and tame; but the soil is light and fertile, being based on sand, gravel, and friable clays, the ancient debris of the formation. On the other hand, the hills of Old Red districts present great diversity of scenery: here rising in rounded heights, there sinking in easy undulations; now swelling in sunny slopes, and anon retiring in winding glens or rounded valley-basins of great beauty and fertility. The Ochils and Sidlaws in Scotland, with their intervening valleys, and the hills of Hereford, Brecknock, and Monmouth in England, belong exclusively to this formation, and may be taken as the type of its physical features.

87. The geographical distribution of the Old Red Sandstone is very extensive, and there are few regions in which one or other of its groups is not clearly developed. In the eastern counties of Scotland all the groups of the system are well exposed; the lower portions occur largely in South Wales and Devon, in the south of Ireland, in Belgium, and in Germany; the middle portions occupy extensive areas in Russia and the flats of Central Europe, in Siberia and Tartary, on to the flanks of the Himalaya Mountains; and different members of the system are found in Central and Southern Africa, in Canada, the United States, and the Brazils. Wherever the system occurs, its strata give ample evidence of oceanic conditions—of broad and tranquil bays, in which were deposited the numerous alternations of the flagstones and tilestones; of sandy shores where the thick beds of sandstones were collected and arranged; and of gravel beaches, which were cemented and solidified into conglomerates and puddingstones. The frequent ripple-marks speak of receding tides, the indentations left by rain-drops tell of heavy showers; while the abundance of corals, crustacea, shell-fish, and fishes, testifies to the exuberance of marine life, in certain areas at least, of the Old

Red Sandstone ocean. And if we turn to the vegetable remains, we find in them, scanty as they may appear, sufficient evidence of marsh, and plain, and hill-side, of rains to nourish, and rivers to transport.

88. Economically, the Old Red system is not of prime importance. From the slaty or laminated beds we obtain such flagstones as those of Arbroath and Caithness, so extensively employed in paving. Greyish building-stone, like that of Dundee and Perth, is obtained from the compacter sandstones of the lower group, and freestone of the finest grain and colour is likewise obtained from the upper group in Roxburgh and in Fifeshire. The felspars, porphyries, and greenstones are exceedingly durable, but are seldom used in building, owing to the difficulty of dressing them into form. They make first-rate road materials, however, and for this purpose are largely employed in the districts where they occur. To the traps of the Old Red the lapidary is chiefly indebted for most of the agates, jaspers, carnelians, and chalcedonies, known as "Scotch pebbles"—these gems being usually found in rough-looking nodules among the debris of the disintegrated rocks, or in the softer amygdaloids which are sometimes quarried for the purpose.

RECAPITULATION.

The system which we have now reviewed under the term of the OLD RED SANDSTONE or DEVONIAN, is one of the most remarkable and clearly defined in the crust of the globe. Characterised on its lower margin by strata containing the remains of fishes, and which form a line of separation, as it were, between it and the underlying Silurian, and defined, on its upper margin, by the rarity of that vegetation which enters so profusely into the composition of the Carboniferous rocks, there can, in general, be no difficulty in determining the limits of the Old Red formation. On the whole, its composition is manifestly arenaceous, the great bulk of the system being made up of sandstones and conglomerates, with subordinate layers of shale and concretionary limestones. Though containing many zoophytes, shells, crustacea, and plant-impressions, its most notable fossils, perhaps, are *fishes*, often of peculiar forms, and all covered either with hard enamelled scales, or with bony plates, and frequently armed with fin-spines. The igneous rocks connected with the system are greenstones, clinkstones, claystones, felspars, porphyries, and other varieties of felspathic traps. These traps are rarely interstratified with the sandstones, and generally appear as disrupting and upheaving masses, either about the commencement or at the close of the period when those hills and ranges were formed, which confer on Old Red districts their peculiarly undulating and diversified scenery. Looking at the whole system, both in point of time and composition, we are prominently reminded of marine conditions —of sea-shores whose sands formed sandstones, and of beaches whose gravel was consolidated into conglomerates and puddingstone — of receding tides that produced ripple-marks, and of showers that left their impressions on the half-dried silt of muddy estuaries. The reddish colour which pervades the whole strata, shows that the waters of deposit must have been largely impregnated with iron—in all probability derived from the earlier granitic and metamorphic rocks whose degradation supplied the sands and gravels of the system. If, on the other hand, we investigate the fossil remains, we are reminded of disturbances which entombed whole shoals of fishes in marine sediment—of marshes and river-banks which gave birth to a scanty growth of ferns, reeds, and rush-like vegetation — and of sedgy margins, where, perhaps, some lowly reptiles enjoyed the necessary conditions of an amphibious existence.

X.

THE CARBONIFEROUS SYSTEM, EMBRACING THE LOWER COAL-MEASURES, MOUNTAIN LIMESTONE, AND TRUE COAL-MEASURES.

89. IMMEDIATELY above the Old Red Sandstone, but clearly distinguishable from it by the abundance of their vegetable remains, occur the lower members of the CARBONIFEROUS SYSTEM. It is to this profusion of vegetable matter—the main solid element of which is carbon—that the system owes its name; a profusion which has formed seams of coal (coal being but a mass of mineralised vegetation), enters into the composition of all the black bituminous or coaly shales, and which stamps many of the sandstones and limestones with a carbonaceous aspect. As above indicated, the system is generally separable into three well-marked groups—*the lower coal-measures*, or *carboniferous slates; the mountain limestone;* and the *upper* or *true coal-measures.* The student must not, however, suppose that these groups are everywhere present one above another in regular order. All that is affirmed by geology is, that these three groups are found in certain localities; and it is a rule of the science always to take as the type of a formation the fullest development that can be discovered. In some districts, as in the north of England, the carboniferous slates are absent, and the mountain limestone with its shales rests immediately on the old red sandstone; in other countries both the lower groups are absent, and the coal reposes on old crystalline rocks; while, on the other hand, in Ireland the carboniferous slates and mountain limestone are enormously developed, and the coal-measures very sparingly and partially so. Whatever portion of the system may be present, it is always easily recognised — the abundance and peculiarity of its fossil vegetables impressing it with features which, once seen, can never be mistaken for those of any other formation. Derived from the waste of all the preceding rocks—the granitic, metamorphic,

silurian, and old red sandstone—the strata of the system necessarily present a great variety and complexity of composition. There are sandstones of every degree of purity, from thick beds composed of white quartz grains, to flaggy strata differing little from sandy shales; shales, from soft laminated clays to dark slaty flags, and from these to beds so bituminous that they are scarcely distinguishable from impure coals; and limestones, from sparkling saccharoid marbles to calcareous grits and shales. Besides these varieties of sandstones, clays, shales, and limestones, there occur, for the first time notably in the crust, seams of *coal* and bands of *ironstone;* and these also appearing in every degree of admixture, add still further to the complexity of the system. On the whole, the carboniferous strata, from first to last, may be said to be composed of frequent alternations of sandstones, shales, limestones, coals, and ironstones—and these in their respective groups we shall now consider.

Lower Coal-Measures, or Carboniferous Slates.

90. This group is meant to embrace all the alternations of strata which lie between the old red sandstone and the mountain or carboniferous limestone. In some districts it is very scantily developed; in others, as in Ireland and Scotland, it attains a thickness of several thousand feet. In the south of Ireland it consists chiefly of dark slaty shales, grits, flaggy limestones, and thin seams of impure coal; and has, from the general slaty aspect of its strata, been termed the *Carboniferous Slates.* In Scotland, particularly in Fife and the Lothians, it has none of this slaty character, but consists essentially of thick-bedded white sandstones, dark bituminous shales, frequently imbedding bands of ironstone, thin seams of coal, and peculiar strata, either of shell-limestone or of argillaceous limestone, thought from its fossils to be of fresh-water or estuary origin. Unless in its fine white sandstones (the ordinary building-stone of Edinburgh and St Andrews), in its fine-grained estuary and shell limestones, and in the greater profusion of its shells and fishes, the lower group, as developed in Scotland, differs little in appearance from the upper group; hence the term *Lower Coal-Measures* generally applied to it in that country.

91. Looking at the lower coal-measures in the mass, there cannot be a doubt they were deposited under very different conditions from the old red sandstone beneath, and the mountain limestone above. Both these formations are eminently marine

—the yellow sandstones being replete with true oceanic fishes, and the mountain limestone profusely charged with marine shells and corals. The lower coal-measures, on the other hand, have more of a fresh-water than of a salt-water aspect. Coralloid fossils are rarely, if ever, found in its strata; its shells are decidedly estuarine; its plants seem to have grown in marine marshes and delta jungles, and many of its fishes are large and of sauroid types.

Lower Jaw and Dentition of Sauroid Fish—Rhizodus Hibberti.

Under these circumstances, we are justified in regarding it as a separate group—a group which, when more minutely investigated, will throw much important light on the earlier history of the period.

92. In its mineral composition and structure, this group bears evidence of frequent alternations of sediment, as if the rivers of transport were now charged with mud and vegetable debris, now with limy silt, and anon with sand and clay. There are no conglomerates as in the Old Red Sandstone, and from the laminated structure of most of the strata, they seem to have been deposited in tranquil waters. There are, however,. more frequent interstratifications of igneous rocks, as if the seas and estuaries of deposit had also been the seats of submarine volcanoes and craters of eruption. The iron which impregnated the waters of the Old Red period, and tinged with rusty red the whole of that system, now appears in the segregated form of thin layers and bands of ironstone. The frequent thin seams of coal point to a new exuberance of terrestrial vegetation, and indicate the existence of a genial climate and of dry lands—of jungles where pines like the araucaria reared their gigantic trunks—of river-banks where tree-ferns waved their feathery fronds—and of estuarine and marine swamps where gigantic reed-like stems, equisetums, and other marsh vegetation, flourished in abundance. When we turn to the shell-limestones, and find them three or four feet in thickness, and entirely composed of mussel-like bivalves, we are instantly reminded of estuaries where these shell-fish lived in beds as do the mussel and other gregarious molluscs of the present

day. Or if we examine the frequent remains of the fishes which are found in the shales and limestones, we have ample evidence of their predaceous habits, and are forcibly reminded of shallow seas and estuaries, where huge sauroid fishes were the tryant-scavengers of the period. A few minute land-shells, and the skeletons of some small reptiles of the frog and lizard kinds, indicate the existence of a terrestrial fauna which becomes more abundant and varied in the higher groups of the system.

Mountain or Carboniferous Limestone.

93. This group is one of the most distinct and unmistakable in the whole crust of the earth. Whether consisting of one thick reef-like bed of limestone, or of many beds with alternating shales and gritty sandstones, its peculiar corals, encrinites, and shells distinguish it at once from all other series of strata. In fact, it forms in the rocky crust a zone, so marked and peculiar, that it becomes a guiding-post, not only to the miner in the carboniferous system, but to the geologist in his researches among other strata. It has received the name of *Mountain Limestone*, because it is very generally found flanking or crowning the trap-hills that intervene between the Old Red and the Coal-Measures, where, from its hard and durable texture, it forms bold escarpments, as in the hills of Derbyshire, Yorkshire, Fife, and many parts of Ireland. It is also termed the *Carboniferous Limestone*, from its occurring in that system, and constituting one of its most remarkable features.

94. As already indicated, this group in some districts consists of a few thick beds of limestone, with subordinate layers of calcareous shale. In other localities the shales predominate, and the limestones occupy a subordinate place, alternating with the shales, thin seams of coal, and strata of gritty sandstones. Occasionally the limestone appears in one bold reef-like mass, of more than a hundred feet in thickness, separated by a few partings of shale, or rather layers of impure limestone. Whatever be the order of succession, it usually occurs as a dark sub-crystalline limestone, occasionally used as marble, but more frequently raised for mortar and agricultural purposes. Along with the other members of the group, it is often replete with the exuviæ of corals, encrinites, and shells, these fossils forming the curious ornamental markings on its polished surface. Besides being rent and dislocated like all other stratified rocks, it is further intersected by what are called *joints* or *divisional planes* (the "backs" and "cutters" of

the quarryman)—these being fissures perpendicular to the lines of bedding, and causing the rock to break up in large tabular masses. These natural rents affording free passage to water, the mountain limestone is very often grooved and channeled; these channels, where the rock is thick, becoming caverns and grottoes of great extent and magnitude. It is to this percolation of water, charged with carbonic acid, that we owe not only these caverns, and the beautiful *stalactites* and *stalagmites* which adorn their roofs and floors, but also the numerous petrifying springs which abound in limestone districts.

95. The fossils of the limestone group are the usual coal-plants in the shales; and in the calcareous beds numerous varieties of corals, corallines, encrinites, shells, trilobites, and enamel-scaled fishes, some of huge size and sauroid aspect. The whole of these fossils are highly indicative of marine conditions, and in

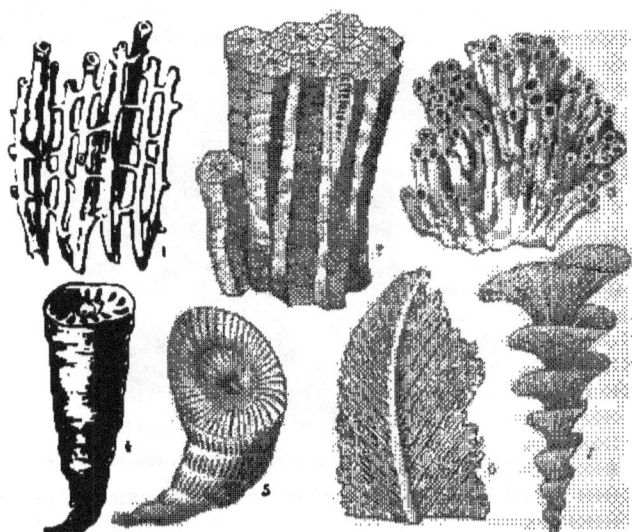

1, Syringopora; 2, Lithostrotion; 3, Aulopora; 4, Amplexus; 5, Clusiophyllum; 6, Fillopora; 7, Archimedopora.

general the observer feels as little difficulty in accounting for the formation of the group, as he does in accounting for the origin of an existing coral-reef. Among the zoophytes the most characteristic are varieties of *retepora* and *flustracea*, whose net-like markings are found in almost every bed of calcareous shale; and of numerous *cup-corals*, *star-corals*, *tube-corals*, and *branching* and *lamelliferous corals*, such as constitute the growing coral-reefs of the Pacific. Of the radiata, by far the most abundant are the *crinoidea* or encrinites, whose jointed stems and branches often make up the entire mass of limestone. As trilobites were especially characteristic of the silurian period, and bony-plated fishes of the old red sandstone, so may encrinites be regarded as peculiarly distinctive of the mountain limestone. They occur in endless varieties, but are all constructed on the same plan—viz., that of a cup-like body, furnished with numerous arms and branches, and attached to the sea-bottom by a jointed and flexible stalk. They derive their names chiefly from the shape of their cup-like bodies, or from that of the calcareous joints which compose the stalk. Thus we have the *cyathocrinite*, so called from the cup-like shape of its body ; the *apiocrinite*, or pear-shaped ; the *pentacrinite*, whose stalk is five-sided instead of round ; the *actinocrinite*, or spiny encrinite ; the *Woodocrinus*, after its discoverer,

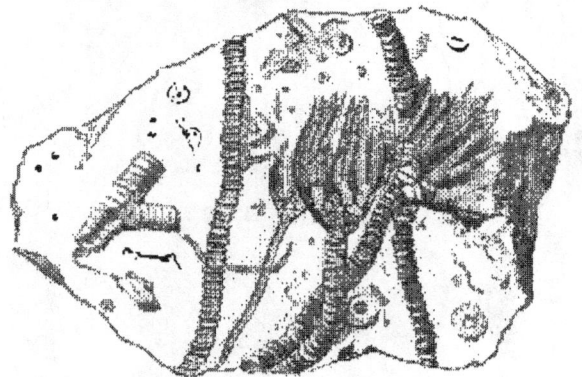

Block of Encrinital Limestone, with Cups and Stems of Woodocrinus macrodactylus.

Mr Wood of Richmond ; and many others, all deriving their names from some marked external character or other circumstance. Besides the encrinites, or lily-shaped radiata, there are true star-fishes, like the *asterias* of our own seas, and *echinoderms*, like our sea-urchins. Of the shell-fish, the bivalves, known as *producta*, *terebratula*, and *spirifera ;* the univalves, *turitella*, *patella*, and *buccinum ;* the coiled chambered-shells, *euomphalus*, *goniatites*, and *bellerophon ;* and the straight chambered-shells known as *orthoceratites*, are the most abundant and characteristic. Of crustaceans, the minute *cypris* and other entomostracan forms abound in myriads; *trilobites* occur in three or four specific

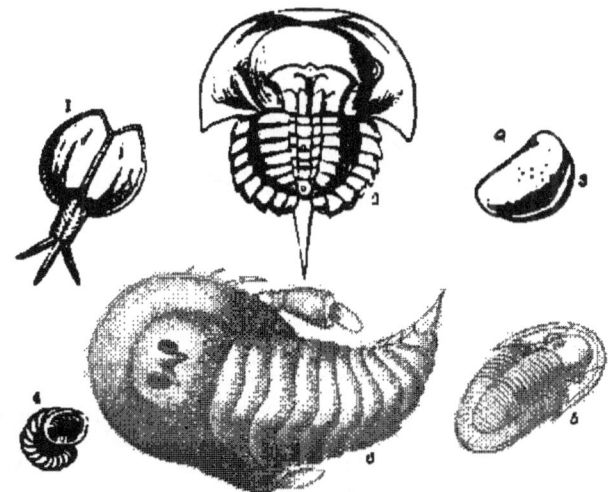

1, Dithyrocaris; 2, Limuloides; 3, Cypris, magnified, and natural size; 4, Spirorbis (Annelid), magnified; 5, Phillipsia (Trilobite); 6, Eurypterus.

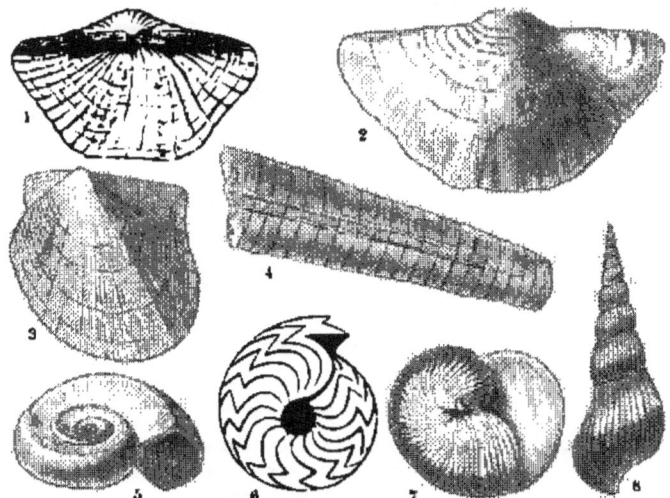

1, Spirifera; 2, Producta; 3, Articulopteron; 4, Orthoceras; 5, Euomphalus; 6, Goniatites; 7, Bellerophon; 8, Loxonema.

forms; and the larger *eurypterites* and *limuloid* forms are by no means uncommon. Of fish-remains, the *holoptychius* (all-wrinkle, from the wrinkled surface of the scales), the *megalichthys* (large

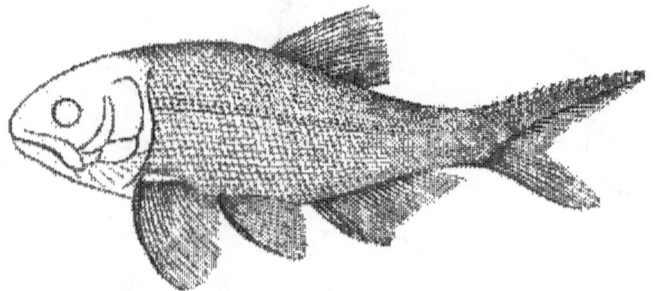

Amblypterus nemopterus.

fish), the *platysomus* (broad shoulder), *palæoniscus, amblypterus,* &c.; the *ichthyodorulites,* or fin-spines of great shark-like fishes, as *gyracanthus* (twisted spine), *ctenacanthus* (comb-spine), &c.; and the palatal teeth of similar generas, as *petalodus, psammodus,*

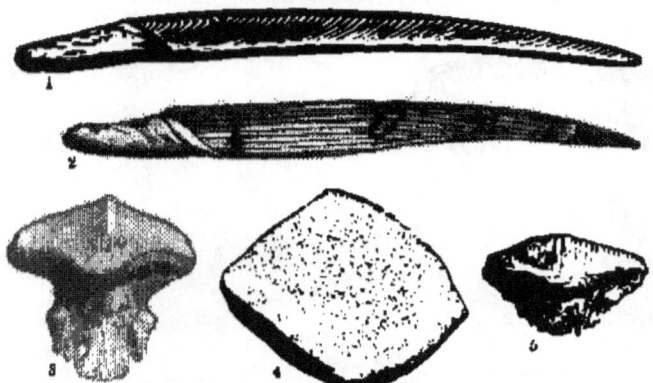

1, Gyracanthus; 2, Ctenacanthus; 3, Petalodus Hastingii; 4, Psammodus porosus; 5, Ctenoptychius serratus.

&c., are perhaps the most abundant in British coal-fields. Besides the remains of sauroid fishes, there occur occasional remains of fish-like saurians, indicating that in the same estuaries and shallow seas, reptiles of aquatic, and, in all likelihood, of amphibious, habits were beginning to prevail. Little is yet known of the anatomy of these early reptiles, but from their fish-like affinities they are generally regarded as of lowly organisation. The coal-fields of Germany, Scotland, Ireland, and Nova Scotia have as yet yielded the most intelligible fragments; and these are known by such names as *dendrerpeton* (tree-reptile), *archegosaurus* (primeval saurian), *anthracosaurus* (coal-saurian), and the like. Another common fossil in the shales of the mountain limestone and coal-measures, as indeed in the shales of all the secondary formations, is the *coprolite* (Gr. *kopros*, dung, and *lithos*, a stone), or fossil excrement of fishes and saurians. In many instances coprolites contain scales, fragments of shells, &c., the remains of creatures on which these voracious animals preyed, and not unfrequently they exhibit the screw-like impression of the intestines.

The Upper Coal-Measures.

96. This group, which completes the Carboniferous system, derives its name from the fact that it furnishes in Britain those valuable beds of coal which contribute so materially to our country's prosperity and power. Occurring immediately above the mountain limestone, or sometimes separated from it, as in the north of England, by thick beds of quartzose sandstone, known as the *Millstone grit*, it consists essentially of alternations of sandstones, grits, fire-clays, black bituminous shales, bands of ironstone, seams of clay, and occasional beds of limestone. One of the most notable features in its composition is the frequent recurrence of seams of coal and beds of bituminous shale—all bespeaking an enormous profusion of vegetable growth, and a long-continued epoch in the world's history when conditions of soil, moisture, and climate conjoined to produce a flora since then unparalleled in rapidity of growth and luxuriance. It is this profusion of vegetable growth, now converted or mineralised into *coal*, which distinguishes the carboniferous from all other systems —the lakes and estuaries of the period being repeatedly choked with vegetable matter, partly drifted from a distance by river inundations, but chiefly and most extensively accumulated on the bed of its growth after the manner of peat-mosses, jungles, and submerged forests.

97. The coal-measures, as already stated, consist of alterna-

tions of sandstones, coals, shales, ironstones, clays, and impure limestones. Among these multifarious beds there is no apparent order of succession, though gritty sandstones may be said to prevail at the base of the group, shales and coals in the middle, and sandstones and marly shales in the upper portion—these gradually passing into the superior system of the new red sandstone. The sandstones occur in great variety, but are in general of a dull-white or brown colour, and thick-bedded. Occasionally they are thin-bedded or flaggy, but in this case they are more or less mingled with carbonaceous, argillaceous, or calcareous matter. The coals also present numerous differences, and are known to mineralogists as *anthracite*, a non-bituminous and semi-lustrous variety; *caking-coal*, a highly bituminiferous sort, like that of Newcastle, which cakes or undergoes a kind of fusion during combustion; *splint*, a less bituminous and slaty variety, which burns free and open, without caking; and *cannel*, a compact lustrous variety, which breaks with a conchoidal or shell-like fracture, and is extensively used in the manufacture of gas. The shales are all dark-coloured, and more or less bituminous; the limestones often impure and earthy; and the ironstones occur in bands or in nodules—either as a clay carbonate of iron ("clay-band") or in combination with bituminous or coaly matter, as the "blackband" of Scotland.

98. The organic remains of the coal-measures, though exhibiting many features in common with the groups already described, are still, as a whole, peculiarly well defined. As an estuary deposit, many of the beds contain shells (the "mussel-bands" or "mussel-binds" of the miner), fishes, and other aquatic exuviæ. A few encrinites appear in certain exceptional beds of limestone, but otherwise marine types are subordinate, and estuary ones prevail. The fishes are chiefly of large size, and of a sauroid character; and the reptiles, as mentioned under the Mountain Limestone (par. 95), show chiefly aquatic characteristics, though in some instances undoubtedly amphibious. In several fields—as those of Germany, Belgium, Nova Scotia, and Britain—we have evidences of terrestrial life in the skeletons of certain lizard and frog-like reptiles, and in fragments of land-shells, and remains of insects; and as research is extended, such remains will no doubt be discovered more abundantly, both in these and other localities. The grand feature of the period, however, is the abundant and gigantic flora, comprising hundreds of forms which have now only distant representatives in tropical swamps and jungles. Araucarian-like pines, tree-ferns, gigantic reeds, equisetums, club-mosses, and other kindred forms, crowd every bed of shale, enter into

many of the sandstones, and constitute solid seams of coal. Of the more characteristic of these forms we may notice the *sigillaria*

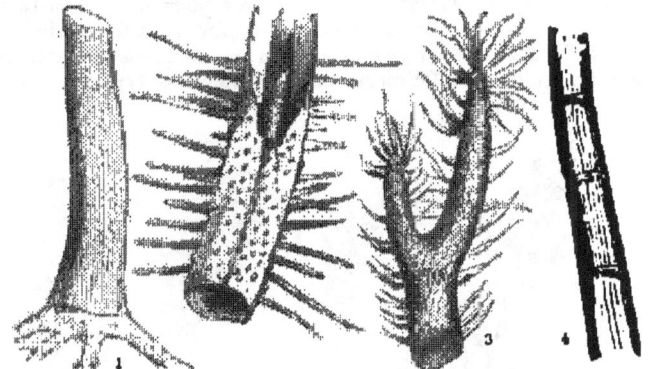

1. Sigillaria; 2. Stigmaria (root of Sigillaria); 3. Lepidodendron; 4. Calamites.

(*sigillum*, a seal), so called from the seal-like impressions on its trunk; the *stigmaria* (*stigma*, a puncture), from the dotted or punctured appearance of its bark, ascertained to be the roots of sigillaria; the *lepidodendron* (*lepis*, a scale, and *dendron*, a tree), from the scaly exterior of its bark; *calamites* (*calamus*, a reed), from the reed-like jointings of its stalk; *hippurites*, from its re-

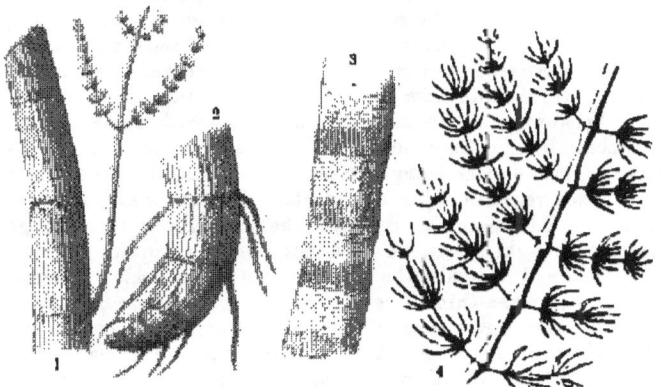

1. Calamites nodosus; 2. Root of Calamite, with rootlets; 3. Hippurites giganteus; 4. Asterophyllites foliosa.

semblance to the *hippuris* or mare's-tail of our marshes; and *asterophyllites* (*astron*, a star, and *phyllon*, a leaf), from the star-like whorls of its leaves. In fact, these and all the other vegetable remains are named from some peculiarity of form, the ablest botanists being yet unable to assign them a place among existing genera. Of the fern-like impressions—so abundant in the shales, and which must meet the eye of the student in almost every fragment he splits—the following may be taken as typical forms:

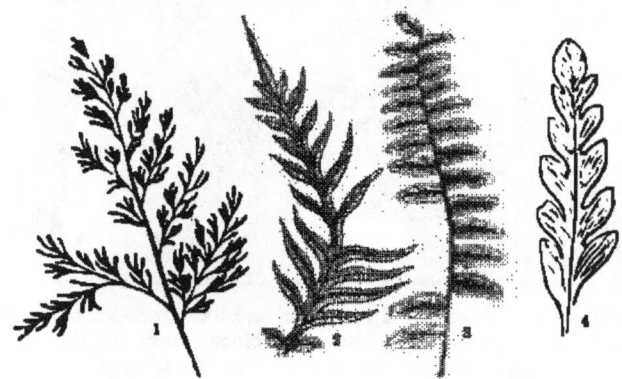

1. Sphenopteris affinis. 2. Pecopteris lonchitica (pinnule); 3. Neuropteris gigantea (pinnule); 4. Odontopteris obtusa (pinnule).

sphenopteris, or wedge-leaf (Gr. *sphen*, a wedge; *pteris*, a fern, from *pteron*, a wing), from the wedge shape of its leaves; *glossopteris*, or tongue-leaf (*glossé*, a tongue); *pecopteris*, or comb-leaf (*pekos*, a comb); *neuropteris*, or nerve-leaf (*neuron*, a nerve); *cyclopteris*, or round-leaf (*kyklos*, a circle); and so on with many similar forms.

99. Whatever the botanical families to which these extinct vegetables belong, they now for the most part constitute solid seams of coal — coal being a mass of compressed, altered, and mineralised vegetation, just as sandstone is consolidated sand, or shale consolidated mud. By what chemical process this change has been brought about, we need not minutely inquire; but we see in peat and in lignite the progressive steps to such a mineralisation; and when thin slices of coal are subjected to the microscope, its organic structure is often as distinctly displayed as the cells and fibres in a piece of timber. Of the amount of vegetation required to form not only one seam, but forty or fifty seams, which often succeed each other in coal-fields, we can form no adequate conception, any more than we can calculate the time required for their growth and consolidation. This only we know, that conditions of soil and moisture and climate must have been exceedingly

favourable; that over a large portion of the globe such conditions prevailed for ages; and that partly by the drift of gigantic rivers, and partly by the successive submergences of forests, jungles, and peat-swamps, the vegetable matter was accumulated which now constitutes our valuable seams of coal.

100. During the whole of the carboniferous epoch we have ample evidence of igneous activity. In the lower coal-measures we have frequent interstratifications of trap-tuff and ash, and these become more abundant in connection with the mountain limestone. Subsequent to the deposition of the system, it seems to have been shattered and broken up by those forces which elevated the trap-hills of the mountain limestone, and gave birth to the numerous basaltic crags and conical heights of our coal-fields. The traps are chiefly augitic, and consist of basalts, greenstones, clinkstones, trap-breccias, trap-tuffs, and earthy amygdaloids. The upheavals and convulsions of the period have greatly dislocated the strata, and most of our coal-fields exhibit trap-dykes, faults, and fissures, in great complexity and abundance. If we except the hills of the mountain limestone, some of the basaltic crags and cones, and now and then a glen of erosion cut through the softer strata of the system, the scenery of coal districts is on the whole rather tame and unpicturesque. The soil, too, in general derived from the shales and clays beneath, is often cold and retentive, and requires all the skill and appliances of modern agriculture to render it moderately fertile. These drawbacks, however, are more than compensated for by the value of the mineral treasures which lie beneath.

101. The industrial importance of the carboniferous system can only be adequately appreciated in a country like Britain, which owes to it the proud mechanical and manufacturing position which she now enjoys. *Building-stone* of the finest quality is obtained from the white sandstones of the lower groups; *limestones* for mortar, hydraulic cement, and agricultural purposes, are largely quarried from the middle group, which also yields *marbles* of no mean quality; *fire-clay* for bricks, tiles, pipes, retorts, &c., is extensively raised from the coal-measures; *ironstone*, both black-band and clay-carbonate, is mined in almost every coal-field, and constitutes almost the sole supply of this metal in Britain; *ochre* (hydrated oxide of iron) is obtained in several localities; *alum* is largely prepared from some of the shales; *copperas*, or sulphate of iron, is manufactured from the pyritous varieties; *paraffine* and *paraffine-oil* from the more bituminous varieties; and our supply of *coal*, in all its varieties, is procured solely from this system. The mountain limestone is also in this country the main reposi-

tory of the ores of *lead, zinc,* and *antimony,* and in the same veinstones are associated ores of *silver,* and not unfrequently of *gold.* On the whole, the carboniferous system is decidedly the most valuable and most important to man; and when we name the principal coal-fields of Britain, we point at the same instant to the busiest centres of our manufacturing and mechanical industry.

RECAPITULATION.

The strata we have now described constitute a well-marked and peculiar system, lying between the Old Red Sandstone beneath, and the New Red Sandstone above. Their most striking peculiarity is the profusion of fossil vegetation, which marks less or more almost every stratum, and which in numerous instances forms thick seams of solid coal. It is to this exuberance of vegetation that the system owes its name—*carbon* being the main solid element of plants and coal. Although this coaly or carbonaceous aspect prevails throughout the whole system, it has been found convenient to arrange it into three groups—the Lower Coal-Measures or Carboniferous Slates, the Mountain or Carboniferous Limestone, and the Upper or True Coal-Measures; or more minutely, as is generally done by British geologists, into—

1. Upper Coal-Measures.
2. Millstone Grit.
3. Mountain Limestone; and
4. Lower Coal-Measures.

Taking the whole succession and alternations of the strata—the sandstones, clays, shales, limestones, ironstones, and coal—and noting their peculiar fossils, the estuarine character of the shells and fishes of the lower and upper groups, and the marine character of the corals, encrinites, shells, and fishes of the middle group, with an excess of terrestrial vegetation throughout, we are reminded of conditions never before or since exhibited on our globe. The frequent alternations of strata, and the great extent of our coalfields, indicate the existence of vast estuaries and inland seas, of gigantic rivers and periodical inundations; the numerous coalseams and bituminous shales clearly bespeak conditions of soil, moisture, and warmth favourable to an exuberant vegetation, and point partly to vegetable drift, and partly to submerged forests, to peat-swamps and jungle-growth; the mountain limestone, with its marine remains, reminds us of low tropical islands, fringed with coral-reefs, and to lagoons thronged with shell-fish and fishes; the existence of reptiles and insects tells us of air, and sunlight, and river-banks; the vast geographical extent of the system bears evidence of a more equable climate over a large portion of the earth's surface; while the interstratified trap-tuffs, the basaltic outbursts, and the numerous faults and fissures, testify to a period of intense igneous activity—to repeated upheavals of sea-bottom and submergences of dry land. All this is so clearly indicated to the investigator of the carboniferous system, that he feels as convinced of their occurrence as if he had stood on the river-bank of the period, and seen the muddy current roll down its burden of vegetable drift; threaded the channels of the estuary, gloomy with the gigantic growth of swamp and jungle; or sailed over the shallow waters of its archipelago, studded with reef-fringed volcanic islands, and dipped his oar into the forests of encrinites that waved below. The natural conditions under which the system was formed are not more wonderful, however, than the economical importance of its products. Building-stone, limestone, marble, fire-clay, alum, copperas, lead, zinc, silver, and, above all, iron and coal, are its principal treasures—conferring new wealth and comfort on the country that possesses them, and giving a fresh and permanent impetus to its industry and civilisation.

XI.

THE PERMIAN SYSTEM, EMBRACING THE MAGNESIAN LIMESTONE AND LOWER NEW RED SANDSTONE.

102. IMMEDIATELY above the Coal-measures—in some instances lying unconformably, and in others insensibly graduating from them—occurs a set of red sandstones, yellowish magnesian limestones, and variegated shales and marls, enclosing irregular masses of rock-salt and gypsum. To this series of strata, as more especially developed in England, the earlier geologists applied the term *New Red Sandstone*, in contradistinction to the old red sandstone system, which we have already described as lying beneath the carboniferous formation. Though the sandstones are not all red, nor the limestones the only magnesian limestones in the crust of the earth, still reddish hues prevail throughout the sandstones and shales as developed in the British Islands, and the calcareous beds are certainly more eminently magnesian than any others with which we are acquainted. At one time the term *Poikilitic* (*poikilos*, variegated) and *Saliferous* (salt-yielding) were applied to the system; but the fact that variegated marls abound in the old red, and that salt is found in several other systems, has rendered these designations inappropriate, and now all but obsolete. This New Red Sandstone of the earlier geologists is now usually arranged into two distinct systems, the *Permian* and the *Triassic*—the former embracing the lower members, which are largely and typically developed in the government of Perm, in Russia; and the latter comprising the upper members known as the "Trias," or triple group, in Germany. The reasons for this arrangement are, that the fossils of the magnesian limestone and lower red sandstones seem more closely allied to those of the coal-measures beneath, than to those of the variegated sandstones and saliferous

marls above; in other words, present a *palæozoic* aspect, while those of the upper sandstones and marls are decidedly *mesozoic*. Indeed, many of the fossils of the Permian or Lower New Red are identical with those of the Carboniferous, and it has been questioned by some whether the so-called "Permian" ought not to be regarded as the termination of the Coal period, rather than a separate and independent system. Following the majority of English geologists, it may be separated in the mean time, but the student must not on that account consider it as a *system* or *life-period*, equivalent, in the proper sense of the term, to that of the Old Red beneath, or that of the Oolite above.

103. The Permian system, as developed in England, Germany, and Russia, consists essentially of reddish and occasionally whitish quartzose sandstones; of reddish and variegated shales (mottled, purple, yellow, and green); of yellowish limestones, containing a notable percentage of magnesia; and of calcareous or marly flagstones, often largely impregnated with copper-pyrites. The sandstones are generally thick-bedded, sometimes gritty, but rarely conglomerate on the large scale, though frequently containing pebbles and intercalated bands of pebbly conglomerate. The shales are usually called "marls," but this less from their containing any notable quantity of lime, than from their occurring in a mottled, friable, and non-laminated state. The limestones vary from an almost pure carbonate of lime to an admixture containing upwards of forty per cent of carbonate of magnesia—hence called "magnesian limestones." Their structure is often peculiar, occurring in thick beds, with subordinate concretionary masses, and layers of a powdery consistence. These concretions are frequently of curious shapes—*honeycombed*, *mammillary* (or pap-like), and *botryoidal* (or in clusters like a bunch of grapes). When the magnesian limestone assumes a granular and crystalline texture, it is known by the mineralogical name of *dolomite*, after the French geologist, M. Dolomieu. The slaty or flaggy beds are known in England as "marl slates," and in Germany, where they are largely impregnated with copper-pyrites, as "keuper-marls," and "kupfer-schiefer" (copper-slate), names now quite familiar to British geologists.

104. With respect to the order of succession among the strata, the Permian, like every other system, presents local differences and irregularities. In England it consists chiefly of red sandstone and grits, of magnesian limestones and gypseous marls, and of laminated calcareous flagstones. This succession is usually tabulated as follows:—

ORDER OF SUCCESSION. 113

MAGNESIAN LIMESTONE.
{ LAMINATED LIMESTONE, with layers of coloured marls, as at Knottingley, Doncaster, &c.
GYPSEOUS MARLS—Red, bluish, and mottled.
MAGNESIAN LIMESTONE — Yellow and white; of various texture and structure; some parts, as at Tynemouth, brecciated, or made up of fragmentary masses.
MARL SLATES—Laminated, impure calcareous flagstones of soft argillaceous or sandy nature.

RED SANDSTONE.
{ RED SANDSTONES, with red and purple marls, and a few micaceous beds. The grits are sometimes white or yellow, and pebbly. When conformable, this sandstone occasionally passes into the coal-measures on which it rests.

In France, Germany, Russia, and North America, where the system has been well investigated, some of these members are wanting, while others are more fully and typically developed. Thus, placing side by side the English and German representatives, we obtain a very complete view of the system, and learn the important fact that it is by general types, and not by any conventional arrangement of strata, that the geologist must be guided in his deductions:—

In England.	*In Germany.*
Laminated limestones.	Stinkstein.
Brecciated limestones.	Rauchwacke.
Fossiliferous limestone.	Dolomit; upper zechstein.
Compact limestone.	Zechstein (mine-stone.)
Marl slate.	Mergel-schiefer and kupfer-schiefer.
Red sandstones and grits.	Rothe-todo-liegende.

105. Taken as a whole, the student must perceive that a great lithological difference exists between the red sandstones, magnesian limestones, and mottled marls of the Permian rocks, and the gritty sandstones, bituminous shales, and coal-seams of the Carboniferous system. On the other hand, when he comes to investigate the fossils, he will find that many forms are common both to the Permian and the Coal-measures; while in the Triassic or upper portion of the New Red, other forms are common to it and the Oolitic beds above. It is for this reason that the Permian or lower new red takes its place among the *Palæozoic* strata, and the Triassic among the *Mesozoic*, as indicated in paragraph 42, to which the student is here requested to refer.

106. The organic remains, as far as discovered within the area of Europe, do not appear to be very abundant, and with this paucity of fossils it would be unsafe to dogmatise too confidently as to the ultimate grouping of all the members of the system. Among the PLANTS specially characteristic of the Permian strata

may be mentioned *sphenopteris, neuropteris,* and other ferns closely allied to those of the coal-measures; *calamites, asterophyllites, lepidodendron, lycopodites* (club-mosses), *equisetums* and *coniferæ,* or pines of the araucarian family Leaves like those of the fan-palms, known by the name of *Nœggerathia,* with silicified

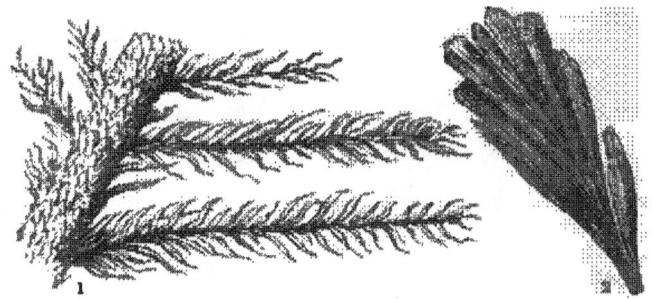

1. Walchia piniformis; 2. Nœggerathia cuneifolia.

trunks of tree-ferns, termed *psaronites,* are common features of the Permian flora. Sigillaria and stigmaria, so eminently characteristic of the carboniferous era, have not yet been detected, and indeed the whole flora seems to be limited and scanty. Of ANIMALS a few *corals* and *corallines* have been found, and

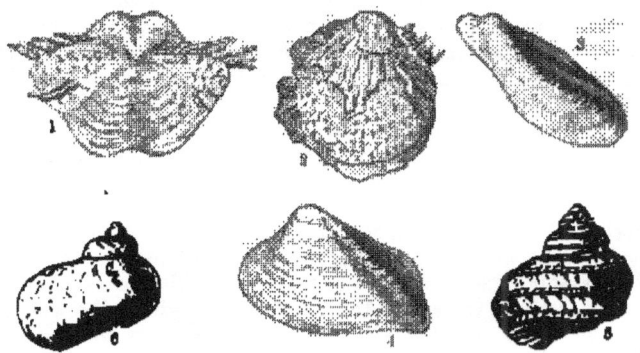

1. Productus horridus; 2. Strophalosia Morrisiana; 3. Bakevellia Sedgwickiana; 4. Schizodus Schlotheimi; 5. Turbo helicinus; 6. Natica Leibnitsiana.

shells like the *productus, strophalosia, trigonotreta, mytilus, schizodus, turbo,* and *natica,* are not uncommon in the magnesian limestone; but we altogether want that profusion of corals, encrinites, and molluscs which thronged the waters of the moun-

ORGANIC REMAINS.

tain limestone epoch. The trilobites have also vanished, and we have only two or three genera (*cythere* and *dithyrocaris*) of minute crustacea. Of fishes we have several of the smaller ganoid forms, as *palæoniscus*, *pygopterus*, and *platysomus*; but with this

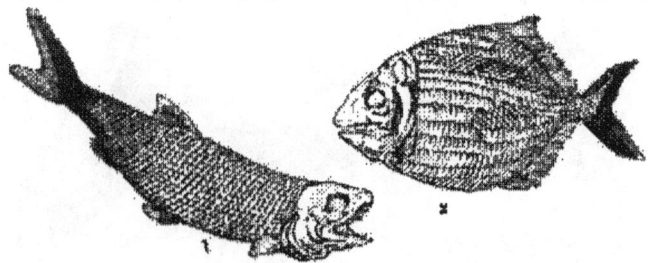

1, Palæoniscus Friesi-beni; 2. Platysomus striatus.

group most of these forms disappear, and are never found in any subsequent formation. Reptile life seems to have been on the increase, and the aquatic and fish-like forms of the coal-measures are succeeded by true air-breathing, land-inhabiting creatures of the frog and lizard families, whose bones, teeth, and footprints are of frequent occurrence throughout the strata of the system. Of these the *palæosaurus* (ancient saurian), *protosaurus* (first saurian), and *thecodontosaurus* (sheath-tooth saurian), are the most common and characteristic—the thecodont saurians being, indeed, as peculiar to the system as the ichthyosaurs are to the lias and oolite.

107. The igneous rocks associated with the strata are chiefly dykes and outbursts of basalt, greenstone, pitchstone, and claystone porphyry. These outbursts seem to be connected with igneous centres situated in the older systems, and pass alike through the old red, carboniferous, and new red systems. With the exception of some tufaceous and brecciated beds at the base of the system, there appear to be no interstratifications of igneous matter; and, on the whole, the Permian era (within the area of Europe) seems to have been one of comparative tranquillity. The consequence is, that districts in which it is the prevailing surface formation are in general flat and tame, being devoid of those eruptive undulations and eminences which give character to the scenery of the mountain limestone and old red sandstone.

108. The industrial products of the system, though not to be compared with those of the coal-measures, are still of considerable

importance. The sandstones are quarried in many districts for building purposes, as are also some of the magnesian limestones (Durham and Yorkshire), which dress well, and are often exceedingly durable. The limestones are likewise used in agriculture, as mortar for the builder, and for the extraction of magnesia; while certain of the compact varieties found in Germany furnish blocks for lithographic printing. Gypsum is an abundant product of some of the marls; while in Germany the *kupfer-schiefer* has been long mined as an ore of copper, and furnishes a large proportion of that valuable metal.

RECAPITULATION.

The system above described consists essentially of reddish sandstones, yellowish magnesian limestones, and slaty calcareous beds. From the prevailing hues of its strata, and from the fact of its lying immediately above the coal-measures, it has been termed the *new red sandstone*, in contradistinction to the old red, which lies beneath. Along with the saliferous marls and variegated sandstones of the triassic strata above, it was early observed to hold a sort of middle place among the secondary formations; hence the lias, oolite, and chalk were considered as *younger* or *upper secondaries*, while the new red, the carboniferous strata, and the old red, were termed the *older* or *lower secondaries*. From the fact of the lower members of the new red sandstone containing fossils more or less allied to carboniferous types, and its upper members imbedding those less or more allied to oolitic forms, it has been separated into two distinct systems—the *Permian* (from Perm in Russia, where the lower beds are extensively developed) and the *Triassic*, regarding the triple group of Germany as typical of the upper strata. Adopting this view, we have the following synopsis:—

TRIAS.	Keuper. Muschelkalk. Bunter sandstein.	Saliferous marls and girts. (*Wanting.*) Variegated sandstone.
PERMIAN.	Lower bunter. Zechstein. Kupfer-schiefer. Rothe-liegende.	Gypseous marls and grits. Magnesian limestone. Marl slate. Red sandstones.

In the Permian the fossils are plants akin to those of the coal-measures, with crinoids, shell-fish, fishes with heterocercal tails, and frog-like reptiles. In the Trias, as will be seen in the next chapter, the plants resemble oolitic types, and the animal remains are corals, encrinites, shell-fish, fishes with homocercal tails, amphibious reptiles, traces of birds, and small marsupial mammals. Taking the whole composition, succession, and remains of both systems, they indicate a period of shallow seas supercharged with saline matter, of muddy estuaries and lagoons, of an arid and warm climate, and of frequent submergences and upheavals. During the period many forms of life disappeared, and were succeeded by others of a different type and order; hence the Permian group is regarded as *palæozoic*, and the Triassic as *mesozoic*. On the whole, the Permian strata have been little disturbed by igneous rocks, and new red sandstone districts are in consequence rather flat and monotonous. The soil is of medium quality, and affords rich verdant pastures rather than arable land for mixed husbandry. Industrially, the system yields building-stone, limestone, gypsum, copper, and occasionally valuable seams of coal, if we adopt the belief of American geologists, that the red and grey sandstones of Virginia and North Carolina are of Permian epoch.

XII.

THE TRIASSIC SYSTEM, COMPRISING THE KEUPER, MUSCHELKALK, AND BUNTER SANDSTEIN OF GERMANY, OR UPPER NEW RED SANDSTONE OF ENGLAND.

100. THE reasons for separating what was formerly known as the "New Red Sandstone" into two distinct systems—the *Permian* and *Triassic*—have been stated in the preceding chapter. Before this division, it was usual to arrange the new red sandstone, as developed in England, into upper, middle, and lower groups—the *upper* comprising the saliferous marls and variegated sandstones of Cheshire; the *middle*, the magnesian limestones of York and Durham; and the *lower*, those reddish sandstones and grits which immediately overlie the coal-measures in the north of England. The succession of the strata composing the lower and middle groups has been tabulated in paragraph 104. The following exhibits the lithology of the upper section :—

VARIEGATED MARLS.—Red, with bluish, greenish, and whitish laminated clays or marls holding *gypsum* generally, and *rock-salt* partially (as in Cheshire). Interstratified with these marls are certain grey and whitish sandstones.

VARIEGATED SANDSTONES.—Red sandstones, with white and mottled portions; the lower strata in some districts pebbly.

In addition to these marls and sandstones, there is developed on the Continent a considerable thickness of shelly fossiliferous limestone known as the MUSCHELKALK; and when this is interpolated, the upper new red consists of three well-marked members; hence the *Trias*, or triple system of German geologists. Tabulated in descending order, the following exhibits the details of the system as developed in Germany and England :—

	Germany.	England.
1. KEUPER.	Saliferous and gypseous shales, with beds of variegated sandstones and carbonaceous laminated clays.	Saliferous and gypseous marls, with grey and whitish sandstones.

	Germany.	England.
2. Muschelkalk	Compact greyish limestone, with beds of dolomite, gypsum, and rock-salt.	(Wanting.)
3. Bunter Sandstein.	Various coloured sandstones, dolomites, and red clays; occasional pisolites.	Reddish sandstones and quartzose conglomerates.

For the purposes of the beginner, it is enough, perhaps, to remember that the Triassic system consists essentially of three groups—1, Keuper marls and grits (saliferous marls); 2, Muschelkalk (shelly limestone); and, 3, Bunter Sandstein (variegated sandstones).

110. When we turn to the fossils of the system we find the Plants of the Coal and Permian epochs represented only by a few *calamites, equisetums*, and ferns, and their place taken by others

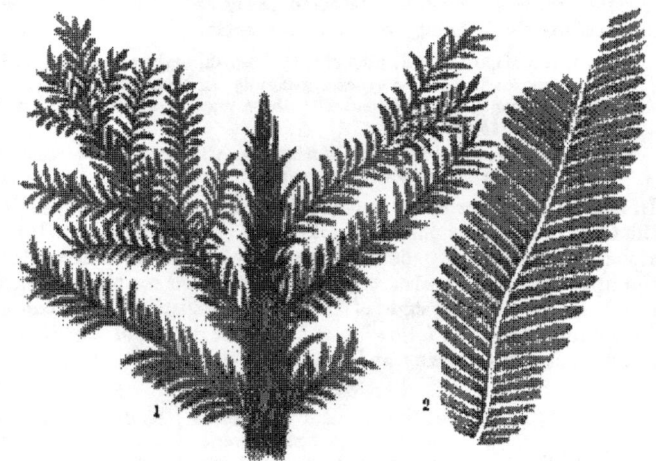

1, Walchia diffusa ; 2, Pterozamites bosaris.

apparently allied to the *palms, cycads, zamias*, and true pines. We have now few corals; but star-fish (*aspidura*) and crinoids (*encrinus liliiformis*) occur abundantly in the German muschelkalk. Of shell-fish, *terebratula* and *spirifer* still occur; but the great flush of brachiopods (*productus*, &c.) which marked the period of the carboniferous limestone is unknown. Of triassic bivalves the *avicula, mya, plagiostoma*, and *ostrea*, are perhaps the most abundant; and of chambered shells the *ceratites*, allied to the ammonite, is the most common genus. Crustaceans of minute forms, *estheria* (posidonia), are abundant in the fine-grained lime-

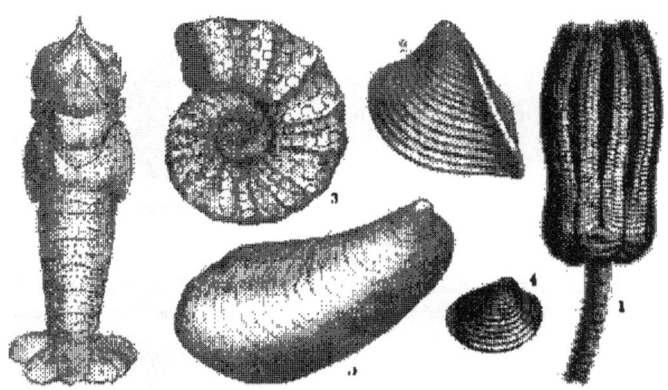

1. Encrinus liliformis; 2. Myophoria lineata; 3. Ceratites nodosus; 4. Estheria minuta; 5. Plagiostoma obliqua; 6. Pemphix Sueurii.

stones; craw-fish like forms (*pemphix*) also occur; and beetle-like insects are by no means uncommon in the same beds. Of sauroid fishes numerous species have been discovered, as the *saurichthys* (sauroid-fish), *gyrolepis* (twisted scale), *acrodus* (pointed-tooth), and others, all deriving their names from some marked peculiarity in appearance. Of reptiles several curious genera have been found, allied to the lizards and monitors of our own time, the most characteristic being the *labyrinthodon* (so named from

Restored outline of Labyrinthodon pachygnathus.—Owen

the structure of its teeth), the *phytosaurus* (plant-saurian), the *nothosaurus* or doubtful saurian, the *rhynchosaurus* (beaked saurian), the small lizard-like *telerpeton* of the Lossiemouth sandstones, and the large crocodilian-like *staganolepis* of the same formation—if that formation shall eventually be proved to be of Triassic age (see par 85). To these may be added the curious *dicynodont*, or two-tusked reptiles from Southern Africa and Bengal, whose jaws, besides being furnished with a cutting edge

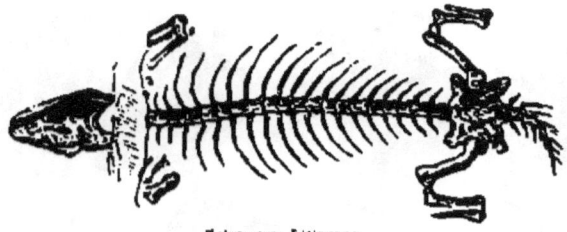

Telerpeton Elginense

like those of the turtles, had a pair of downward-curving tusks in the upper jaw like those of the walrus. Besides the teeth and

1, Rhynchosaurus articeps; a, vertebra of do.; 2, Dicynodon lacerticeps; 3, Skull of Sudanolepis.

bones of these early reptiles, we have also their footprints impressed and preserved on the slabs of sandstone, almost as clearly as if they had traversed the muddy beach of yesterday. These footprints speak a language similar to that of the ripple-mark and the rain-drop formerly alluded to—the foot leaving its impress on the yielding and half-dried mud, and the next deposit of sediment filling up the mould. On splitting up many of these slabs of sandstone, the mould and its cast are found in great perfection —so much so, that not only the joints of the toes but the very texture of the skin is apparent. These fossil footprints, termed *ichnites* (from *ichnon*, a footstep), have been largely found at Corncockle Muir in Dumfriesshire, at Cummingstone in Morayshire (the Lossiemouth sandstones), at Storeton in Cheshire, at Hildburghausen in Germany, on the Connecticut in America, and many other places. Some of them are evidently reptilian, hence termed *sauroidichnites;* others, again, appear to be those of gigantic birds, and thence termed *ornithichnites* (*ornis*, a bird); while others appear to be those of unknown quadrupeds (in all likelihood of some huge batrachian or frog-like reptile), and have received the provisional designation of *tetrapodichnites*, or four-footed imprints. The annexed engraving represents the footsteps of the *cheirotherium* (*cheir*, the hand), so called from the hand-like impressions of its feet, and is supposed by Professor Owen to

Footprints of Cheirotherium.

be one and the same with the batrachian or frog-like labyrinthodon. Still higher in the scale of being than either reptiles or birds—whose bones (*palæornis*) as well as footsteps have recently been found—are the mammalian teeth, discovered several years ago in the bone-breccia of Wurtemberg, a stratum which occurs among the upper beds of the Keuper. These remains are those of a warm-blooded quadruped—the earliest of its kind then detected in the crust of the earth; and from this circumstance, and their diminutive size, the creature received the provisional name of *microlestes antiquus* (Gr. *mikros*, little; *lestes*, a beast of prey). Analogous remains (*dromatherium silvestre*) have also been de-

Jaw of Dromatherium silvestre, from the Red Sandstones of North Carolina.—EMMONS.

tected in the Red Sandstones of Virginia and North Carolina—strata which by some are regarded as Triassic, and by others as the equivalents of our European Permians. And more recently still (1860), mammalian jaws and numerous detached teeth have been discovered in the Trias near Bristol, thus adding important links to that gradually lengthening chain which is yearly connecting the Past more intimately with the Present, and leading geologists to dogmatise less rashly regarding the creational introduction and progress of vitality.

111. Strata, as above described—that is, strata composed of reddish clays and marls, containing deposits of rock-salt and gypsum, of greyish shelly limestones, and variegated sandstones, and pebbly grits—are found occupying considerable areas both in the Old and New Worlds. They occur in patches on the western coasts and islands of Scotland; on the shores of Morayshire; in the basin of the Solway; on the opposite coast of Ireland; and a broad belt of the same strata runs across the whole of England, from Durham to Devon on the one hand, and to Lancashire on the other. On the Continent it occupies still wider areas in France, Germany, and the region of the Alps, while some of its most instructive features are exhibited in the southern states of North America.

112. The igneous rocks associated with the system are identical with those which break through and displace the Permian strata. On the whole, triassic districts are little varied by trap eruptions, while the predominance of clays and shales and soft sandstones give rise to broad level expanses, rather tame and uninteresting in their superficial features. "Spread over so immense a space in England," says Professor Philips, "the triassic system offers the remarkable fact of never rising to elevations much above 800 feet—a circumstance probably not explicable by the mere eroding of these soft rocks by floods of water, but due to some law of physical geology yet unexplained. We only can conjecture that it is connected with the repose of subterranean forces which prevailed after the violent commotions of the coal strata, over nearly all Europe, till the tertiary epoch." In general the new red sandstone districts are better fitted for pasture and dairy husbandry than for the purposes of mixed husbandry or corn-culture.

113. Reviewing the whole new red system (Permian and Triassic inclusive)—its sandstones, shales, magnesian limestones, gypseous, saliferous, and cupriferous marls, its comparatively few plants, its marine shells and fishes, its reptiles and fossil footprints, and its generally flat and undisturbed position—we are reminded of quiet shallow seas, of iron-tinged rivers, and of estuaries studded with lagoons and mud-banks. The marl and copper slates give evidence of tranquil deposit; the footprints, of mudbanks baked and dried in the sun, over which birds and reptiles traversed till the next return of the waters; the gypsum, rocksalt, and magnesia, of highly saline waters, subjected to long-con-

tinued evaporations, or at least to some chemical conditions favourable to the precipitation of these abundant salts; and the presence of iron, colouring less or more the whole strata, together with copper in many of the slates, point to impregnations by no means favourable to the exuberance of marine life. The remains of arborescent ferns and palm-like stems, together with the skeletons and tracks of huge lizard-like reptiles, and remains of small marsupials, bespeak an arid rather than a genial climate, and a want of those conditions which give birth to the exuberant vegetation of the coal era.

114. The industrial products yielded by the system are sandstones of various quality, calcareous flagstones, limestone, gypsum, and rock-salt. Our chief supply of salt, formerly obtained by evaporation of sea-water, is now procured from the salt-mines and brine-springs of Cheshire and Worcester. "The Cheshire deposits of salt lie along the line of the valley of the Weaver, in small patches, about Northwich. There are two beds of rock-salt lying beneath 126 feet of coloured marls, in which no traces of animal or vegetable fossils occur. The upper bed of salt is 75 feet thick; it is separated from the lower one by 30 feet of coloured marls, similar to the general cover; and the lower bed of salt is above 100 feet thick, but has nowhere been perforated. Whether any other beds lie beneath those is at present unknown. They extend into an irregular oval area, about a mile and a half in length, by three-quarters of a mile in breadth." The salt rock in these deposits—as likewise at Middlesborough on the Tees, in Antrim, at Wurtemberg in Germany, and at Vic and Dieuze in France—is sometimes pure and transparent, and at other times is of a dirty reddish hue, and mixed to the amount of half its bulk with earthy impurities.

RECAPITULATION.

The system described in the preceding paragraphs consists essentially of saliferous marls, shelly limestones, and variegated red and whitish quartzose sandstones. Briefly tabulated, it exhibits in England and on the Continent the following series:—

Germany.	*England.*
Keuper marls and grits.	Saliferous marls and grits.
Muschelkalk.	(*Wanting.*)
Bunter Sandstein.	Variegated sandstones.

It has received its name, Trias, or Triassic, from being composed of the three members so clearly developed in Germany, while the name "Upper New Red" is sufficiently distinctive of its place and character among English strata. Its fossils are all of *Mesozoic* types; and though a few point to Permian analogues, the identities, if identities there be, are to be sought among Oolitic rather than among Palæozoic strata. Though the accumulation of such masses of rock-salt (chloride of sodium) be still in some measure an unsolved problem, their occurrence in conjunction with gypsum (sulphate of lime) and with magnesian limestones (carbonates of magnesia and lime) less or more throughout the entire new red sandstone system, would seem to indicate peculiar marine conditions—conditions of shallow bays and lagoons, periodical isolation of certain areas, and again their submergence, and reception of deposits of ferruginous mud and clay-silt. The nature of the imbedded plants and animals points to a warm and arid climate — to a region like Australia, where small marsupials tenanted the scrubby plains—and to shallow estuaries, lagoons, and mud-banks, where wading birds and amphibious reptiles found subsistence on shell-fish, crustacea, star-fishes, and fishes, and left their tracks on the sun-baked mud as evidence of their forms, and of the kind of life they led on the shores of these primeval waters.

XIII.

THE OOLITIC SYSTEM, COMPRISING THE LIAS, THE OOLITE, AND WEALDEN GROUPS.

115. WE have now passed the boundary of the older rocks, and enter upon the upper or younger secondary formations. Or, speaking palæontologically, we have traced the history of systems whose fossils were all of *Palæozoic* types, and now proceed to interpret the records of those that are unmistakably *Mesozoic*. The curious graptolites and trilobites that crowded the Silurian seas have vanished, the bone-cased pterichthys and coccosteus of the Old Red Sandstone have died away, and the sigillariæ and lepidodendra that thronged the jungles of the coal period are now no more repeated. Their places are taken by other forms of plants and animals—forms still widely different from existing races, yet more akin to them than were those of the palæozoic epochs. The Triassic group, as already stated, is considered as marking the dawn of this new cycle of being, which we shall afterwards find closes with the cretaceous or chalk system. In thus attaching high importance to fossils as exponents of the past conditions of the world, lithological and physical distinctions must not be disregarded. There are facts frequently brought to light, and truths explained, by the composition, structure, and relation of rocks, which no profusion of fossils could ever interpret; and here the student is reminded that, however attractive palæontological discoveries may be, they are only of true geological value when taken in connection with chemical, mineral, and mechanical characteristics.

116. The system which we are about to describe consists, as developed in England, of three well-marked groups — the Lias, the Oolite, and the Wealden. Indeed, so clearly defined are these groups that they are sometimes regarded as independent systems; and were it not for certain fossil as well as lithological resemblances that pervade them, this course would in many

respects be preferable. As it is, the *Oolitic system* comprehends the whole of those peculiar limestones, calcareous sandstones, marls, shales, and clays which lie between the new red sandstone beneath and the chalk formation above. And however similar these strata may be in some features, there is no truth in geology more fully established than this, that where the system is complete, the argillaceous laminated limestones and shales termed the *Lias* constitute the lowest group; the yellowish oolitic limestones, calcareous sandstones, sand and clays called the *Oolite*, the middle group; and the greyish laminated clays, with subordinate layers of limestone and flaggy sandstones, the *Wealden* or upper group. Taking these groups in order, the following synopsis exhibits their character as typically developed in England:—

WEALDEN. {
  WEALD CLAY.—Greyish laminated clays imbedding concretions of ironstone, thin layers of argillaceous limestone, and sandy flags.
  HASTINGS SANDS.—Sands and sandstones frequently ferruginous; beds of clay and sandy shale more or less calcareous.
}

OOLITE. {
  PURBECK BEDS.—Estuarine limestones alternating with sands and clays (formerly grouped with the Wealden).
  UPPER OOLITE.—Coarse and fine grained oolitic limestones, with layers of calcareous sand (*Portland stone and sand*); dark laminated clays, with gypsum and bituminous shale (*Kimmeridge clay*).
  MIDDLE OOLITE.—Coarse-grained, shelly, and coralline oolite, with calcareous sands and grits (*Coral-rag*); dark-blue clays, with subordinate clayey limestones and bituminous shale (*Oxford clay*).
  LOWER OOLITE.—Coarse, rubbly, and shelly limestones (*Cornbrash*); laminated shelly limestones and grits (*Forest marble*); thick-bedded oolite, more or less compact and sandy (*Bath, or Great oolite*); flaggy grits and oolites (*Stonesfield slate*); fuller's earth and clay, calcareous freestone, and yellow sand (*Inferior oolite*).
}

LIAS. {
  UPPER LIAS.—Thick beds of dark bituminous shale; beds of pyritous clay and alum shale; indurated marls or marlstone.
  LOWER LIAS.—Dark laminated limestones and clays; bands of ironstone; layers of jet and lignite; beds of calcareous sandstone.
}

117. It will be perceived from the preceding synopsis that the *Lias* or *liassic* group occupies the lowest portion of the system, and that it is essentially composed of dark argillaceous limestones, bluish clays and bituminous and pyritous shales. The name *lias*, which is said to be a provincial corruption of the word

layers, refers to the thin beds in which its limestones usually occur. "The peculiar aspect," says Sir Charles Lyell, "which is most characteristic of the lias in England, France, and Germany, is an alternation of thin beds of blue or grey limestone, with a light-brown weathered surface, separated by dark-coloured argillaceous partings; so that the quarries of this rock, at a distance, assume a striped and ribbon-like appearance." Once seen, this banded appearance of a lias cliff is not easily forgotten; but it must be remembered that the clays generally predominate, and that they contain occasional layers of jet or other coal (*jet* being but a lustrous variety of coal), and bands of ironstone nodules or *septaria*. Most of the shales are bituminous and pyritous, and it is not uncommon, after wet weather, for the Yorkshire cliffs, which are composed of these beds, to ignite spontaneously, and burn for several months. Besides *pyrites* (sulphuret of iron), these shales are impregnated with sulphates of magnesia and soda, with salt (chloride of sodium), and other saline compounds which indicate a marine origin. Indeed, the whole aspects of the lias—its fossils, composition, and lamination—are those of a tranquil deep-sea deposit.

118. The *Oolite*, as a group, consists of more frequent alternations, and is more varied in its composition than the lias. It derives its name from the rounded concretionary grains which compose many of its limestones—these grains resembling the roe or egg of a fish (*oon*, an egg, and *lithos*, a stone). Oolite is the general term, though many of its limestones are not oolitic; *roestone* is sometimes employed when the grains are very distinct; and *pisolite*, or peastone (*pisum*, a pea), when the grains are large and pea-like. As a series, the oolite consists of alternations of oolitic limestones, calcareous grits, shelly conglomerates, yellowish sands, and thick-bedded, bluish-grey clays less or more calcareous. The peculiar roe-like grains which constitute the oolitic texture, consist either entirely of lime, or of an external coating of lime collected round minute particles of sand, shells, coral, &c.; the grits are composed of fragments of shells, coral, and sand; and many of the strata have a brecciated aspect, hence known as *ragstones*. Like the lias, the oolite is strictly a marine deposit, but its corals, broken shells, and grits, point to shallower waters, to tidal beaches, and sandbanks.

119. The *Wealden group*—so termed from the "Wolds" or "Wealds" of Kent and Sussex, where the deposit prevails—consists chiefly of clays and shales, with subordinate beds of indurated sands, sandstones, and shelly limestones that indicate an estuarine or brackish-water origin. Thin partings of lignite and

bituminous shale are not unfrequent among the clayey strata. The group is of limited extent in England, and in many countries is altogether wanting; the chalk in such cases resting immediately on the oolite. As typically developed in Kent and Sussex, the wealden seems to occupy the site of an ancient estuary, which received the clay and mud of some gigantic river, whose waters occasionally bore down the spoils of land plants and land animals, to be entombed along with those of aquatic origin.

120. The organic remains of the system, as already stated, are all *Mesozoic*—that is, belong to genera and species differing from those found in the older rocks, and differing also, though less in general aspect, from those of the tertiary and present epochs. They are exceedingly numerous and well preserved, and have long and intimately engaged the attention of palæontologists. VEGETABLE REMAINS are frequent in all the groups, and frequently in such profusion as to form seams of lignite, jet, and coal. The Kimmeridge bituminous shale, known as "Kim coal," the carbonaceous shales, lignites, and coals of eastern Yorkshire, the coal of Brora in Sutherlandshire, as well as the coals of Richmond in Virginia, India, and the Indian islands, belong to the oolitic members of the system. Some of the marine deposits contain impressions of *fuci* or sea-weeds; and in those of estuarine origin,

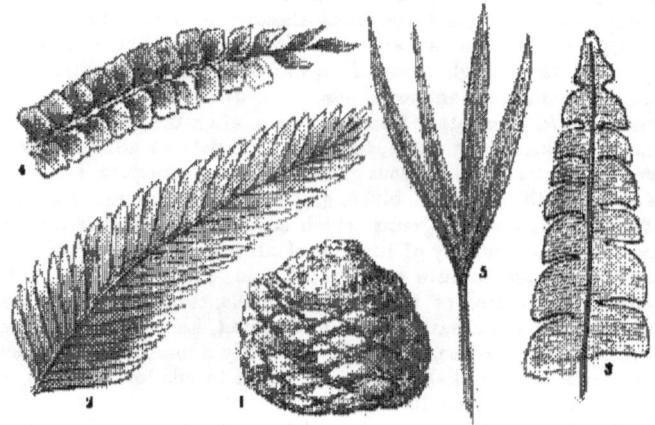

1, Mantellia udiformis; 2, Pterophyllum comptum; 3, Zamites intermedius. 4, Cyclopteris Beanii; 5, Glossopteris elegans.

equisetums, calamites, and other aquatic plants are not uncommon. The terrestrial and marsh orders seem to indicate a genial, or subtropical, climate—the more characteristic forms being arborescent ferns (*pecopteris, cyclopteris*, &c.); palms (*palmacites*), like the pandanus, or screw-pine; cycadeæ closely resembling the existing cycas and zamia, termed *cycadites* and *zamites*, with their fern-like leaves *pterophyllum* and *pterozamites;* coniferæ apparently allied to the araucaria, yew, and cypress, and known as *thuytes, strobilites, peucites,* &c.; and abundance of monocotyledonous leaves resembling those of the lily, pine-apple, agavè, aloe, and allied genera. One of the most remarkable facts connected with the vegetation of the period, is the occurrence of dark loam-like strata, locally known as the "dirt-beds" of Portland, and which must have formed the soils on which grew the cycas and other oolitic plants, though now interstratified with limestones, sandstones, and shales. "At the distance of two feet," says Mr Bakewell, "we find an entire change from marine strata to strata once supporting terrestrial plants; and should any doubt arise respecting the original place and position of these plants, there is over the lower dirt-bed a stratum of fresh-water limestone, and upon this a thick dirt-bed, containing not only cycadeæ, but stumps of trees from three to seven feet in height, in an erect position, with their roots extending beneath them. Stems of trees are found prostrate upon the same stratum, some of them from twenty to twenty-five feet in length, and from one to two feet in diameter."

121. With respect to the ANIMAL REMAINS, we have representatives of almost every existing order, with the exception of the higher mammalia. Beginning with the lowest forms, we have *spongia* or sponges; numerous zoophytes, more like the corals of existing seas than those of the mountain and silurian limestones, and among which the more common forms are *astræa* (star-corals), *madrepores, millepores, meandrina* (brain-corals), and *turbinolia*, a variety of cup-in-cup coral; crinoids, of which the *apiocrinite* and *pentacrinite* are the most frequent; star-fishes, like the *asterias* and *ophiura;* sea-urchins, like the *echinus, clypeus*, and beautiful *cidaris;* worm-like annelida, as *serpula* and *vermicularia;* insects in the Stonesfield slate and Lias like the cricket and dragon-fly —*gryllus* and *libellula;* and crustacea, like the minute bivalved *cyprides,* the cray-fish-like *eryon* and *megacheirus*, and the woodlouse, *archæoniscus*. Of shell-fish, as might have been expected in a system so eminently marine, there is a vast profusion belonging to every order; and of these we may notice the more characteristic forms—viz., the cockle-like bivalves, *cardium, isocardia,* and *trigonia;* the borers, *pholadomya* and *pholas;* the mussel-like *modiola;* the oysters, *ostrea* and *gryphæa;* the clams, *pecten*

and *plagiostoma* ; the whorled univalves, *nerinæa* and *pleurotomaria* ; and, above all, the chambered shells, *ammonite, nautilus,*

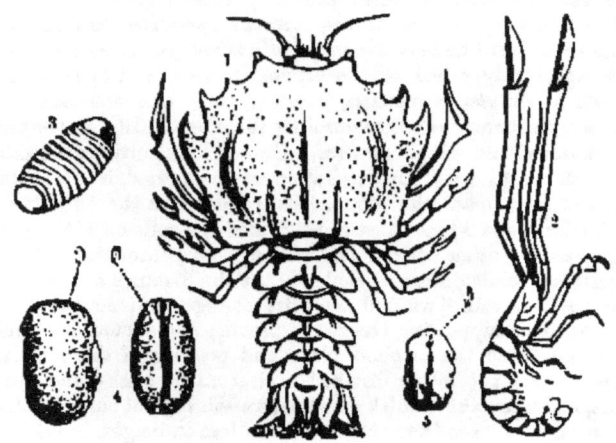

1, Eryon; 2, Megachairus; 3, Archæoniscus; 4, 5, Cyprides—natural size, and magnified.

and *belemnite*. Of these testacea, the gryphæa is so abundant in the lias, that it is sometimes termed the "gryphite limestone,"

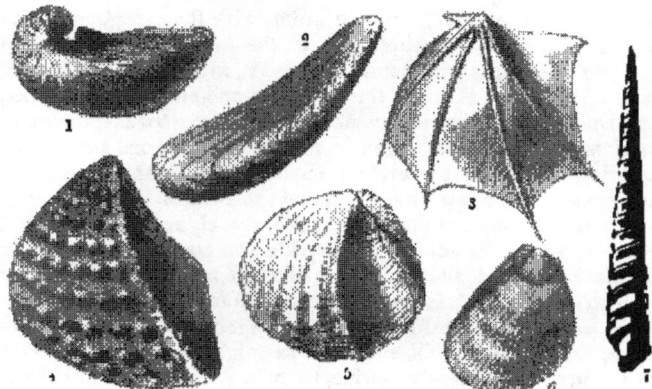

1, Gryphæa; 2, Modiola; 3, Avicula; 4, Trigonia; 5, Pholadomya; 6, Plagiostoma.
7, Nerinæa.

and for a similar reason one of the Jura oolites is called the "neringan limestone." The most notable order of mollusca belonging to the period was undoubtedly the *cephalapodous*—that is, having their organs of motion arranged round the head, like the nautilus and cuttle-fish. Of these the ammonite (so called from its resemblance to the curved horn on the head of Jupiter Ammon) seems to have thronged the waters in many hundreds of

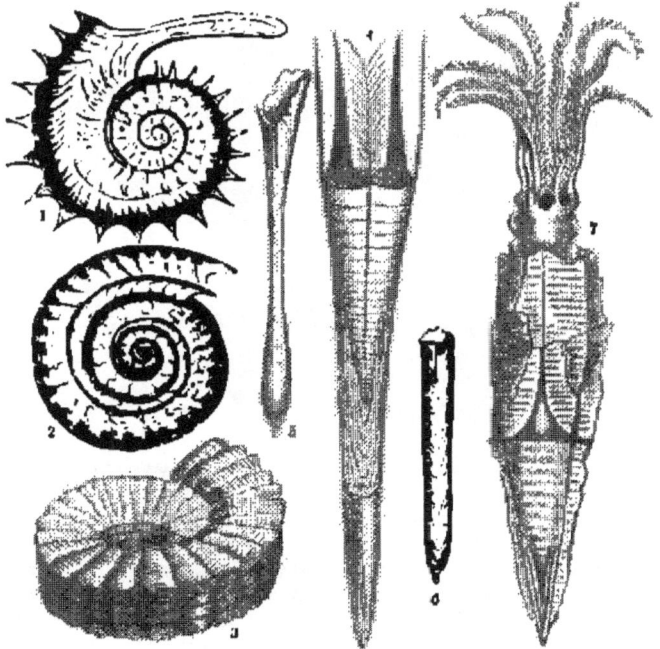

1, Ammonites Jason; 2, A. communis; 3, A. Bucklandi; 4, Belemnites Puzosianus. 5, 6, Belemnites; 7, Belemnoteuthis.

species, and of all sizes, from shells of half an inch to shells of three feet in diameter. Gigantic cuttle-fishes were also the congeners of the ammonite and nautilus, and have left evidences of their existence in the *belemnites* (*belemnos*, a dart), which were the internal bones of these marvellous mollusca.

122. Of the higher or Vertebrated forms of life we have many examples of placoid and ganoid fishes, of saurcid reptiles, and four or five species of marsupial mammals. Of the ichthyolites, the teeth, *hybodus* and *acrodus*, resemble those of the shark-like

cestracion now inhabiting the Australian seas; the fin-spines are often nearly a foot in length, and serrated on one or both sides; and the large enamelled scales of the *lepidotus, tetragonolepis,*

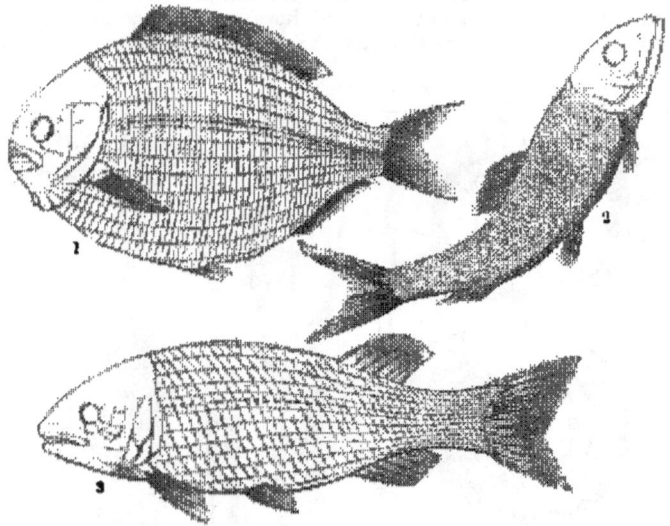

1. Dapedius tetragonolepis; 2. Leptolepis spratt.formis. 3. Lepidotus Valdensis.

and others, bear testimony to the size of these estuarine and marine fishes. Of the reptiles there are several forms of tortoise and turtle (*chelonia* and *platemys*); and others seem distinctly allied to the crocodiles, gavials, monitors, and iguanidons of tropical climates, but differing widely in their structure and apparent modes of existence. One of the most frequent forms is the *ichthyosaurus* (*ichthys*, a fish, and *saurus*, a lizard), somewhat

Ichthyosaurus communis

resembling the crocodile, but furnished with paddles or flippers instead of limbs. Many species have been discovered, and hundreds of individuals varying in length from four to forty feet.

Another common form is the *plesiosaurus* (so called from its greater resemblance to the lizard tribe), distinguished by its enormous length of neck, smaller head, and shorter body and

Plesiosaurus dolichodeirus.

tail. A third and frequent form is the *pterodactyle* (*pteron*, a wing, and *daktylos*, a finger), so called from being furnished with membranous wings, and capable, like the bats, of mounting in

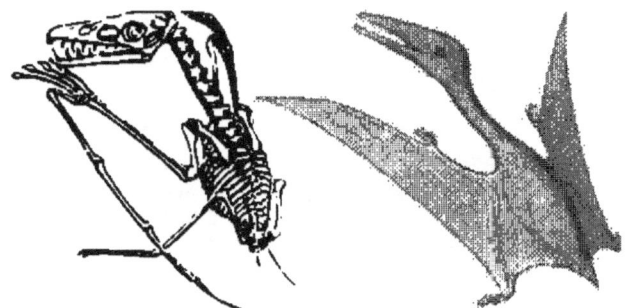

Pterodactylus crassirostris.

the air. Besides these, many other species of saurians have been figured and described by geologists—some aquatic, and others amphibious or terrestrial ; some carnivorous, and others evidently herbivorous. They are generally known by such names as *cetiosaurus* (whale-like saurian), *megalosaurus* (great saurian), *geosaurus* (land saurian), *hylæosaurus* (forest saurian), *teleosaurus*, (perfect saurian), and *iguanodon*, from the almost perfect identity of the teeth and skeleton of a huge fossil reptile found in the wealden with those of the living iguana of America. Bird-remains, so rare in every formation, have recently been discovered in the lithographic limestones of Germany. These, at first regarded as belonging to some species of pterodactyle, have been shown by Professor Owen and others to be those of a true bird, having (unlike any living bird) a long vertebrated tail, but in other respects analogous in bone and feather. This unique specimen, now in the British Museum, has been named *archæopteryx* or "ancient feather-wing," and strengthens the opinion that bird-life existed during the earlier periods of the Trias and Permian.

Archæopteryx macurus; Tail and detached bones.

Of mammals we have evidence in certain jaw-bones, teeth, and detached bones found in the flaggy limestones of Stonesfield. These are ascribed to marsupial animals allied to the opossum and kangaroo of Australia, and have been named *amphitherium* (doubtful) and *phascolotherium* (*phaskolos*, a pouch). More re-

Oolitic Mammals, natural size—1. Lower Jaw and Teeth of Phascolotherium. 2. of Triconodon; 3. of Amphitherium; 4. of Plagiaulax.

cently, and in particular during the summer of 1857, numerous specimens of teeth and jaws and detached bones were discovered in the middle Purbecks of Dorsetshire—some of them insectivorous, others herbivorous, and all, with one or two exceptions, belonging to small marsupial quadrupeds. The discovery, therefore, of the mole-like *spalacotherium* (*spalax*, a mole), in 1854, and of the hoofed hog-like *stereognathus* (solid-jaw), during the same year, have, in 1857, been followed by that of the *triconodon* (three-coned tooth)—the *plagiaulax* (*plagiaulacodon*, oblique-grooved tooth), and others—thus again correcting the hasty generalisations of limited observation, and pointing the warning finger to those who would attempt to dogmatise on the imperfect data which Geology has yet at its command.

123. The igneous rocks associated with the oolitic system in England are gentle outbursts of trap and intersecting dykes of greenstone. There appear to be no contemporaneous trap effusions, and, on the whole, the system retains much of its original sedimentary flatness. In the north and west of Scotland, and in the Jura mountains, it is upheaved and disrupted by granitic rocks, but these may be regarded as subordinate to the trappean compounds which occur in those ranges. The physical features of oolitic districts are by no means unpleasing—the alternations of limestones and clays on a grand scale producing a succession of rounded ridges and sloping valleys. These undulations are very marked in some districts of England and France, where the limestones, which have resisted denudation, compose the ridges, and the softer clays and shales the valleys. None of these ridges are of great height, and being on a limestone subsoil, are dry and fertile, and present a marked contrast to the stiff soils of the "coombs" and "wolds" occupied by the lias and wealden clays. The areas overspread by the oolitic system are rather limited and partial. It is most typically developed in England, where it occupies a broad strip stretching from Yorkshire to Dorset; detached patches occur in the north and west of Scotland; and portions of the system are found in Germany, Switzerland, and France, where the oolitic members are generally known as the "Jurassic system." It is found skirting the Apennines in Italy; flanking the southern Himalayas, and in Cutch in India; in the Malayan peninsula, and Indian archipelago; and recently equivalent beds, with workable seams of coal, have been detected near Richmond in Virginia—coal-seams, which by some American geologists are regarded as of Triassic and Permian formation.

124. Respecting the conditions of the world during the deposition of the wealden, oolite, and liassic strata, we have already stated that everything reminds us of a genial and equable climate. "The close approximation of the amphitherium and phascolotherium," says Professor Owen, "to marsupial genera now confined to New South Wales and Van Diemen's Land, leads us to reflect upon the interesting correspondence between other or-

ganic remains of the British oolite and other existing forms now confined to the Australian continent and adjoining seas. Here, for example, swims the *cestracion*, which has given the key to the nature of the palates from our oolite, now recognised as the *teeth* of congeneric gigantic forms of cartilaginous fishes. Not only *trigoniæ*, but living *terebratulæ* exist, and the latter abundantly, in the Australian seas, yielding food to the cestracion as their extinct analogues doubtless did to the allied cartilaginous fishes called *acrodi* and *psammodi*, &c. Araucariæ and cycadeous plants likewise flourish on the Australian continent, where marsupial quadrupeds abound, and thus appear to complete a picture of an ancient condition of the earth's surface, which has been superseded in our hemisphere by other strata, and a higher type of mammalian organisation."

125. Industrially, the system is by no means devoid of importance. Some of the oolite sandstones, like those of Bath and Portland, form excellent building-stones, and are largely used in the metropolis; while paving-stones and tile-stones are obtained from the indurated flags of the wealden. The limestones of the lias and oolite are largely quarried for mortar; those of the former, when well prepared, furnishing an excellent hydraulic cement. Marbles of various quality are procured from the lower beds of the weald, in Sussex, and also from some of the coralline and shelly oolites, as at Whichwood Forest in Oxfordshire, whence the term "forest marble." Fuller's earth, at one time extensively used in woollen manufacture, is a product of the oolite, and alum is obtained from the lias shales of Yorkshire. Seams of coal, which are often workable, occur in the oolite, as in Yorkshire, at Brora in Sutherlandshire, at several places in Germany, near Richmond in Virginia, and Chatham in North Carolina, in India, the Indian islands, and other localities. Indeed, many foreign coal-fields are now known to be of oolitic origin, or at all events to be of Mesozoic or Secondary age, and later than the true Carboniferous era. Jet, which is only a compact variety of coal, and lignite or wood-coal, are both found in the system, though neither is of much economic value. And recently the ironstone of Cleveland in Yorkshire, which occurs in thick beds, and over a large extent of that county, has added new industrial importance to the oolite, and wealth to that district of England—the busy population of northern Yorkshire and Middlesborough being the direct result of this discovery.

RECAPITULATION.

The Oolitic system, as typically developed in England, is separable into three well-marked groups—the Lias, the Oolite, and Wealden. So distinct in many respects are these groups, that they are sometimes treated as independent systems, and in all likelihood the progress of discovery in other countries will soon establish the necessity of some such division. As it is, we have adopted the usual grouping, which may be briefly tabulated as follows :—

WEALDEN.	Weald clays. Hastings sands.	
OOLITIC or JURASSIC.	Purbeck beds (formerly grouped with the Wealden). Portland stone and sand, Kimmeridge clay,	Upper.
	Coral rag, Oxford clay,	Middle.
	Cornbrash and forest marble, Bath or great oolite, Stonesfield slate, Fuller's earth and clay, Inferior oolite,	Lower.
LIASSIC.	Lias clays and marlstone. Lias limestones and shales.	

From the preceding synopsis, it will be seen that the system is mainly composed of argillaceous limestones, limestones of oolitic texture, calcareous sandstones, shelly and coralline grits, clays and pyritous shales, with layers of coal, jet, and lignite. All the members are well developed in England; but it is chiefly the liassic and oolitic that are found in France, Switzerland, and Germany; and patches of the oolite in Scotland, in Hindustan, and in North America. As deposits, the lias and oolites are eminently oceanic, but the lower members of the wealden appear to be marine, while its upper are evidently of estuarine or freshwater formation. With the exception of the higher mammalia, almost every existing order is represented in the fauna of the oolite, but the forms are all Mesozoic, and died out at the close of the chalk era. The vegetation of the system is also extremely varied, but the highest orders appear to be conifers and palms—no example of a true exogenous timber-tree having yet been detected. Of its numerous fossils, the most characteristic are the *cycadaceæ*, of which the stems, fruits, and leaves are found in abundance; the shells of the *gryphæa*, so peculiarly plentiful in the lias; the coiled-up *ammonites* of innumerable species; the *pterodactyle*, or flying lizard; the fresh-water and marine *turtles*; and above all, the *ichthyosaurus*, *plesiosaurus*, and other sauroid reptiles, whose marvellous forms and variety have suggested for the oolite the not inappropriate title of "the age of reptiles." Still higher in the scale of being than these are the remains of birds fitted for aerial flight, *archæopteryx*, and the marsupial mammals, *amphitherium*, *phascolotherium*, *spalacotherium*, *plagiaulax*, &c., and the small-hoofed mammal, *stereognathus*, described and figured by Mr Charlesworth and Professor Owen. Building-stone, paving-stone, limestone, marble, alum, coal, ironstone, jet, and lithographic slate, are the principal economic products of the system.

XIV.

THE CHALK OR CRETACEOUS SYSTEM, COMPRISING THE GREENSAND AND CHALK GROUPS.

126. IMMEDIATELY above the fresh-water beds of the Wealden in the south of England occur a set of well-defined marine sands, dark marl-clays, and thick beds of chalk. These strata, which seldom exceed in the aggregate 1000 or 1500 feet in thickness, constitute the *Cretaceous system*—chalk (L. *creta*) being the most prominent and remarkable feature in the formation. Though neither of great thickness nor widely developed as to area, the Chalk is in many respects one of the most remarkable rock-systems in the British Islands, and has consequently long attracted the research of geologists. As the uppermost member of the younger secondaries, it closes the record of Mesozoic life; and of the innumerable species which composed the flora and fauna of the secondary epochs, comparatively few have been detected in tertiary or post-tertiary strata. Lithologically, it is composed of cretaceous, argillaceous, and arenaceous rocks—the first predominating in the upper, and the two last in the lower portion of the system. The strata, as occurring in the south of England, are usually grouped as follows:

CHALK.
- UPPER CHALK.—Generally soft white chalk, containing numerous flint and chert nodules more or less arranged in layers.
- LOWER CHALK.—Harder and less white than the upper, and generally with fewer flints. (Reddish in the north of England, and with abundance of flints.)
- CHALK MARL.—A greyish earthy or yellowish marly chalk, sometimes indurated.

GREENSAND.
- UPPER GREENSAND.—Beds of silicious sand, occasionally indurated to chalky or cherty sandstone, of a green or greyish white, with nodules of chert.
- GAULT.—A provincial name for a bluish tenacious clay, sometimes marly, with indurated argillaceous concretions and layers of greensand.
- LOWER GREENSAND.—Beds of green or ferruginous sands, with layers of chert and indurated sandstones, local beds of gault, rocks of chalky or cherty limestone (Kentish rag), and fuller's earth.

ORGANIC REMAINS.

127. The preceding synopsis affords a sufficient outline of the composition and succession of the cretaceous system. Of course, considerable local differences occur, and it is sometimes difficult to determine the equivalents of the beds as typically developed in Kent and adjoining counties. Thus, the lower chalk of Yorkshire and of Havre in France contains abundant flint nodules; in Devon and Dorset a gritty bed with numerous fossils occurs towards the base of the chalk; in Lincoln and York a stratum of red chalk is thought to represent the gault of the southern counties; and the Kentish ragstone, which is largely quarried near Maidstone, is wholly unrepresented in the Isle of Wight. When we come to compare the Continental strata with those of England, still wider differences prevail; and in North America the rocks which are charged with cretaceous fossils are often mere sands and clays, sometimes even shingly, and only in certain districts associated with the beds of yellow coralline and silicious limestones. The lower greensand is sometimes termed the "Neocomian group" (*Neocomiensis*, rock of Neufchâtel), this portion of the system being thought to be more typically developed in the neighbourhood of Neufchâtel in Switzerland; but recent facts scarcely support this view, and for all practical purposes the terms Chalk, Gault, and Greensand are sufficiently distinctive.

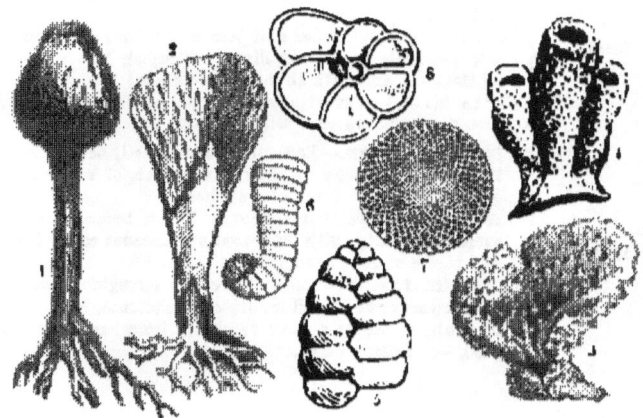

1. Siphonia; 2. Ventriculites; 3. Manon; 4. Scyphia; 5. Textularia; 6. Lituola; 7. Orbitoides; 8. Rotalia.

128. The organic remains found in the cretaceous system are, with a few exceptions, eminently marine, comprising fucoids,

sponges, corals, star-fishes, molluscs, crustaces, fishes, and reptiles. As might be expected, FOSSIL PLANTS are comparatively rare in the British chalk-rocks, and these for the most part drifted and imperfect fragments; but among the cretaceous beds of other regions, seams of lignite and coal are by no means unfrequent. The marine species are apparently allied to the algæ, confervæ, &c., and are termed *chondrites* and *confervites*. The terrestrial types are fragments of tree-ferns, cones of coniferous trees, cycadites and zamites, and are known by such names as *pinites*, *strobilites* (*strobilus*, a fir-cone), *carpolithes* (*carpos*, a fruit), and *samiostrobus*. Of the ANIMAL remains, which are in general beautifully preserved and to be seen in almost every collection, we can only notice one or two examples under each order or family. Of spongiform bodies we have the common and characteristic *choanites*, *spongia*, *scyphia*, and *ventriculites*. Of zoophytes and polyzoa there are numerous *astrea*, *alveolites*, and *orbitolites*, *flustra* and *retepora*. Of echinoderms or sea-urchins there are many species in every state of perfection, as *cidaris*, *galerites*,

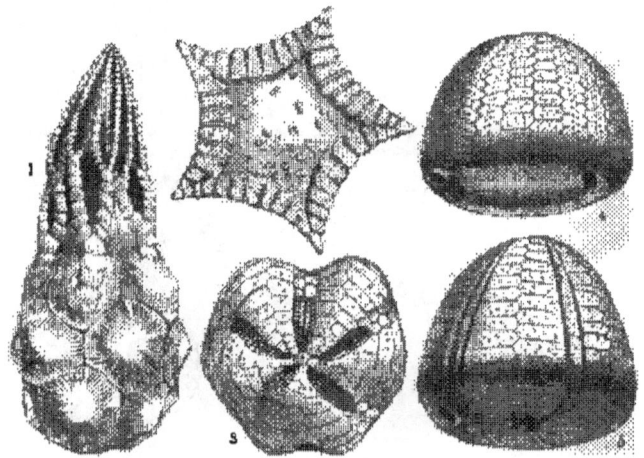

1, Marsupites. 2 Goniaster; 3. Hemipneustes; 4. Ananchytes. 5. Galerites.

spatangus, and *micraster*. Of foraminiferous shells, which compose in great part the chalk strata, there are *rotalia*, *dentalina*, *textularia*, &c. Of annelids, abundant *serpularia* and *vermicularia*; and of crustaceans, species of lobsters, *astacus*, and of crabs, *dagurus*. The remains of testaces or shell-fish are extremely numerous, and in such a state of perfection that the conchologist can at once assign them a place in his classification. Of the char-

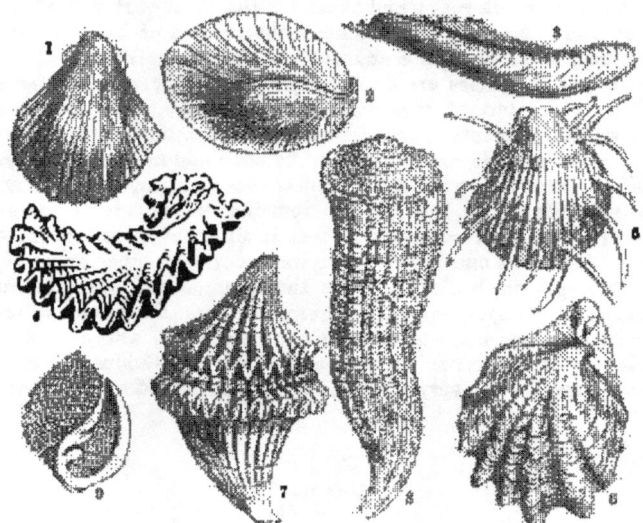

1, Pecten; 2, Terebratula; 3, Gervillia; 4, Ostrea; 5, Plagiostoma; 6, Inoceramus; 7, Radiolites; 8, Hippurites; 9, Cinulia.

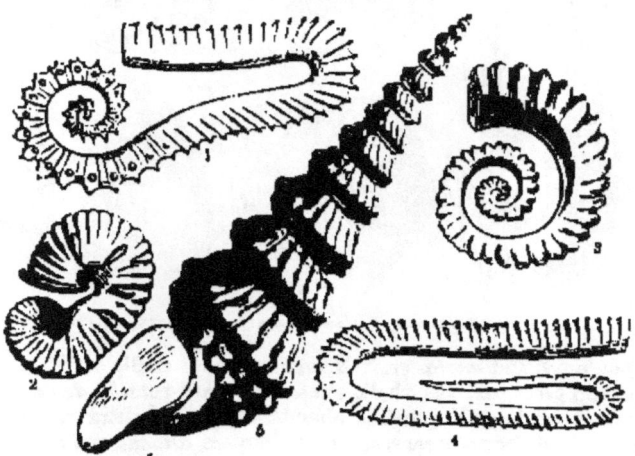

1, Ancyloceras; 2, Scaphites; 3, Crioceras; 4, Hamites; 5, Turrilites.

acteristic bivalves may be noticed *arca, cardium, trigonia, gryphæa, inoceramus, ostrea, pecten,* and *terebratula,* with the curious massive shells *hippurites* and *diceras.* Among the univalves or gasteropods, *cerithium, rostellaria, dentalium,* and *natica,* are typical and characteristic. The chambered shells also appear in vast profusion, and in highly curious forms, as compared with those of the earlier formation. Of these the coiled-up *ammonite,* the dart-like *belemnite* (the "thunderbolts" of the English peasant), the hook-shaped *hamite* (*hamus,* a hook), the boat-shaped *scaphite* (*scapha,* a skiff), the rod-like *baculite* (*baculus,* a staff), and the *nautilus,* are the most frequent and typical.

129. The vertebrate remains are those of fishes and reptiles, with occasional remains of birds and mammalia. Of the fishes the majority are still placoid and ganoid; but the ctenoid and cycloid orders, to which almost all existing fishes belong, are here

1. Corax pristodontus; 2. Lamna crassidens; 3. Otodus obliquus; 4. Lamna elegans; 5. Notidanus microdon.

for the first time found in the rocky strata. Of the placoids the teeth and spines are as usual the only remains; the former being

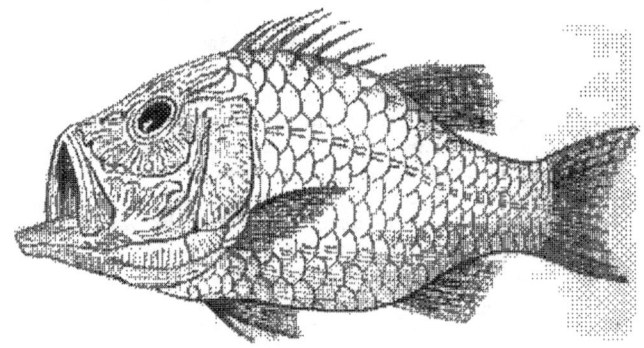

Beryx Lewesiensis.

most abundantly represented by *acrodus, ptychodus* (wrinkletooth), *otodus, corax, lamna,* &c., and the latter by *chimæra,*

Of the ganoids, *lepidotus, gyrodus* (twisted-tooth), and *pycnodus* (thick-tooth), are the most typical. Of the ctenoid or comb-scaled order several species of *beryx* (closely allied to the perch) have been detected; and of the cycloideans the *saurocephalus* and *osmeroides* are those most frequently found in collections. The sauroid reptiles seem identical with, or at least closely allied to, those of the wealden, and are represented by *pterodactylus, plesiosaurus, mososaurus, iguanodon*, and *chelonia*. Bird-bones, termed *cimoliornis* (Gr. *kimolia*, white chalk-marl, and *ornis*, a bird), have been described by Professor Owen, who has also surmised that certain mammalian remains are those of quadrumana or monkeys.

130. Regarding the geographical distribution of the chalk, though the several areas may be partial or limited, strata containing the peculiar fossils of the system have been discovered in many countries. As already mentioned, it is fully developed in the south and south-east of England; it is found in the north of Ireland; and from the frequent occurrence of flint nodules (as in the Buchan district of Aberdeenshire) it is supposed to have covered at one time the oolites of the north of Scotland. It is spread over wide areas in France and Germany, and occurs in connection with the Alps, Carpathians, and Pyrenees. Chalk fossils have been collected in the south of India; equivalents of the gault and greensands have been investigated in the states of New Jersey, Texas, and Alabama; in the north-western prairies, and Vancouver Island; and strata apparently of the same age have been noticed in Colombia in South America.

131. Though exhibiting faults and fractures, no igneous rocks have been found associated with the chalk of England. In the north of Ireland the strata are indurated, disrupted, and overlaid by basalt and other traps, as remarkably displayed at the Giant's Causeway; and in the Pyrenees and Alps the system partakes more or less of all those upheavals, by traps and secondary granites, which are so characteristic of these lofty ranges. Where unbroken by igneous eruptions, the physical aspect of chalk districts is readily distinguished by the rounded outlines of their hills and valleys, as typically exhibited in the "wolds" and "downs" of Kent and Sussex. These downs are described as "covered with a sweet short herbage, forming excellent sheep-pasture, generally bare of trees, and singularly dry even in the valleys, which for miles wind and receive complicated branches, all descending in a regular slope, yet are frequently left entirely dry; and, what is more singular, contain no channel, and but little other circumstantial proof of the action of water, by which they were certainly excavated."

132. Combining all the features of the system—its composition, fossils, and geographical distribution — we are warranted in regarding the chalk as a truly marine deposit, filling up limited sea-areas which were thronged with oceanic life, and which received at intervals the drift of rivers that flowed through countries enjoying a high and genial temperature. The cycas and zamia are plants which betoken a warm climate; and though vegetable drift seldom appears among the chalk strata of Europe in such profusion as to form more than scattered patches of lignite (as in the lower measures near Rochelle), yet must this circumstance be ascribed more to the unfavourable position of the sea of deposit for the reception of such drift, than to the scantiness of vegetation on the dry land. Among the cretaceous strata of the Saskatchewan prairies and Vancouver Island, there occur workable seams of coal—thus showing that the chalk period, like every other, had its areas of exuberant and coal-forming vegetation. Again, the corals and huge sauroid reptiles betoken more of tropical than of temperate conditions, a circumstance that seems further established by the presence of remains apparently allied to the monkeys. Respecting the conditions of the waters in which the chalk, so unlike ordinary limestones, was deposited, and within whose mass flints were subsequently aggregated, geologists are by no means agreed, though the calcareous ooze now forming over so vast an area of the North Atlantic seems to point the way to a satisfactory explanation. This much, however, is certain, that the chalk is a mechanical deposit, from waters loaded with calcareous particles, and abounding in minute forminiferous shells which constitute a large portion of the mass, and not, as at one time supposed, a precipitate from chemical solution. The abundance of enclosed sponges, corals, shells, and fragments of vegetables, also confirms this view, and compels us to seek for the enclosed layers and nodules of flint an origin similar to that of nodules of ironstone and chert in shale. Flints are composed almost entirely of pure silex, with a trace of iron, clay, and lime; they are usually aggregated round some nucleus of sponge, shell, sea-urchin, or other organism; and there is no difficulty in conceiving the silex to have been originally in solution in the waters of deposit, and subsequently segregated in layers and nodules as we now behold it.

133. Industrially, the chief products of the system in Britain are chalk and flint. Chalk, as an almost pure carbonate of lime, is calcined like ordinary limestones, and employed by the brick-layer, plasterer, cement-maker, and farmer; and levigated, it furnishes the well-known "whiting" of the painter. Flint calcined

RECAPITULATION. 147

and ground is used in the manufacture of china, porcelain, and flint-glass; and before the invention of percussion-caps, was in universal use for gun-flints. In the south of England flints are employed as road-material; and the larger nodules are sometimes taken for the building of walls and fences. Beds of fuller's earth occur in the lower series; and some of the indurated strata, like the "Kentish rag" and Chalk-marl of Cambridgeshire, furnish local supplies of building-stone. From the Gault and Lower Greensand are also obtained the coprolitic nodules, now ground down and used as a manure, on account of their containing a large percentage of phosphate of lime. More recently, several of the workable coal-seams of Vancouver Island and British North America have been discovered to belong to the cretaceous epoch.

RECAPITULATION.

The cretaceous system—so called from the chalk-beds which form its most notable feature—is the last or uppermost of the secondary formations. All its types of life are strictly Mesozoic, and of the numerous species found in the Trias, Oolite, and Chalk, comparatively few have been detected in Tertiary strata. It is an error, however, to suppose that there is any sharp line of demarcation either between Palæozoic and Mesozoic, or Mesozoic and Cainozoic. As one rock-system runs into another system, so does one great Life-period pass into another period, and it is only when viewed as a whole that we can note the differences that exist between their respective plants and animals. Our "groups," and "systems," and "periods" are mere provisional expedients; useful as such, but leading to error when invested with any other significance. As typically developed in the south of England, the Cretaceous system has been separated into two groups, the *Chalk* and *Greensand*, and these comprise, in descending order, the following members:—

CHALK.
 { Upper chalk with flints.
 { Lower chalk without flints.
 { Chalk marl.

GREENSAND.
 { Upper greensand.
 { Gault.
 { Lower greensand.

These various members, both in point of composition and fossil remains, bear evidence of deposit in seas of limited area, and of a climate suitable for the growth of cycadaceæ on land, and of corals and gigantic saurians and turtles in the waters. Palæontologically, its remains are eminently marine, and comprise numerous species of sponges, corals, star-fishes, sea-urchins, shell-fish, crustacea, fishes, and reptiles. Remains of birds and mammalia have also been detected, but these are too imperfect and obscure to warrant any definite conclusion as to their character and affinities. The chief, and indeed the only, industrial products of the system in England are chalk, flint, phosphate nodules, and some inferior building-stones; but in some foreign localities lignite, coal, and ironstone occur in available abundance.

XV.

THE TERTIARY SYSTEM, EMBRACING THE EOCENE, MIOCENE, PLIOCENE, AND PLEISTOCENE GROUPS.

134. THE earlier geologists, in dividing the stratified crust into primary, secondary, and tertiary formations, regarded as tertiary all that occurs above the Chalk. The term is still retained, but the progress of discovery has rendered it necessary to restrict and modify its meaning. Even yet the limits of the system may be said to be undetermined—some embracing under the term all that lies between the Chalk and Boulder-drift, others including the drift and every other accumulation in which no trace of man or his works can be detected. Palæontologically speaking, much might be said in favour of both views, but the difficulty of unravelling the relations of many clays, sands, and gravels, makes it safer to adopt in the mean time a somewhat provisional arrangement. We shall therefore treat as TERTIARY all that occurs above the chalk to the close of the drift, and as POST-TERTIARY every accumulation which appears to have been formed since that period. Adopting this arrangement, we have in England the following intelligible subdivisions:—

POST-TERTIARY. { RECENT and SUPERFICIAL ACCUMULATIONS occurring above the boulder-drift.

TERTIARY. { PLEISTOCENE......Boulder-drift.
PLIOCENE ....Mammalliferous and Red crag.
MIOCENE............Coralline crag.
EOCENE.............Strata of London and Hampshire basins.

By adopting this view we get rid of certain anomalies connected with the boulder-drift, while there will be no difficulty in removing the pleistocene to the post-tertiary system, should subsequent discoveries render such a transposition necessary.

135. The organic types of the system above indicated are all *Cainozoic*—that is, are all more or less allied to, or even identical

with, many existing genera. As at the close of the *Palæozoic*, cycle graptolites, trilobites, eurypterites, pterichthys, coccosteus, cephalaspis, megalichthys, sigillaria, lepidodendron, and other forms of ancient life had passed away, so, at the close of the *Mesozoic*, the ichthyosaurus, plesiosaurus, pterodactyle, palæoniscus, ammonites, and encrinites disappeared, and their place was taken by other and more recent-like forms. We now find among vegetables evidence of true exogenous timber-trees (that is, trees which increase by *external* layers of growth, like the oak, beech, and elm); a large percentage of the corals and shells are identical with those of existing seas; the reptiles are carapaced turtles, tortoises, and crocodiles; the fishes are chiefly ctenoids and cycloids with equally-lobed tails; birds of existing families are by no means rare; and examples of mammalia of all orders up to the highest, save man, have been detected. It is true that certain genera and species discovered in tertiary strata are not to be found beyond the limits of the pliocene group; and it is this extinction of many peculiar forms that warrants the separation of the post-tertiary from the tertiary system. The groups are founded upon this perceptible approach to existing species—taking the fossil shells as the index. Thus, *eocene* (Gr. *eos*, the dawn, and *kainos*, recent) implies that the strata of this group contain only a small proportion of living species, which may be regarded as indicating the dawn of existing things; *miocene* (*meion*, less) implies that the proportion of recent shells is less than that of extinct; *pliocene*, (*pleion*, more), that the proportion of recent shells is more or greater than that of the extinct; and *pleistocene* (*pleiston*, most), that the shells of this group are mostly those of species inhabiting the present seas.

Eocene, Miocene, and Pliocene Groups.

136. We arrange these groups under one head, because they are all evidently sedimentary deposits resulting from the ordinary operations of water, and because they are all less or more fossiliferous, and thus give evidence of the geographical conditions of the world during the period of their formation. The case is different with the pleistocene or boulder-drift group, which is clearly not an ordinary sedimentary deposit, and which, with rare exceptions, is altogether destitute of fossils. Confining our remarks to the three lower groups, we find the composition and succession of their strata so extremely varied and irregular, that it is next to impossible to give anything like a generally applicable descrip-

ORDER OF SUCCESSION.

tion. This much may be said, that their areas are usually well defined, as if originally deposited in inland seas or estuaries; that they give evidence of frequent alternations of marine with fresh-water sediments; and, on the whole, are less consolidated than the rocks of older systems. They consist for the greater part of clays, sands, and gravels, with interstratified limestones, calcareous grits, marls, and occasional layers of lignite. With respect to the composition and succession of their strata, the following synopsis of the English tertiaries will convey a better idea than any detailed description:—

PLIOCENE.
- MAMMALIFEROUS CRAG of Norfolk and Suffolk.—Consisting of shelly beds of sand, laminated clay, and yellowish loam, with layers of flinty shingle reposing on the chalk, and generally covered with a thick bed of gravel.
- RED CRAG of Norfolk and Suffolk.—A deep ferruginous shelly sand and loam, with an abundance of marine shells, frequently rolled and comminuted.

MIOCENE.
- CORALLINE CRAG.—A mass of shells and corals in calcareous sand; or compact, and forming flaggy beds of limestone, with bands of greenish marl. Some of the harder portions are used as building-stone.

EOCENE.
- FLUVIO-MARINE BEDS of Hampshire and Isle of Wight. —Consisting of clays and marls sometimes indurated, of sandy clays and subordinate layers of silicious limestone.
- BAGSHOT SANDS.—A series of loose sands, sandstone, greenish sandy clay, and fissile marls.
- LONDON CLAY.—A brown or dark-blue or blackish tenacious clay, with layers of argillo-calcareous nodules. Layers of greenish sand, and masses of gypsum, and iron pyrites not unfrequent.
- BOGNOR BEDS.—Occur towards the base of the London clay, and consist of calcareous and silicious nodules, or of coarse, green indurated sand, with numerous marine shells.
- PLASTIC CLAY AND SANDS.—Composed of sand, shingle, mottled clays, and loam, with beds of rolled flints and marine shells.

The above may be taken as sufficiently descriptive of the English tertiaries; and for the sake of comparison, we subjoin a section (in descending order) of the strata in the Paris basin, which are usually regarded as the equivalents of the English eocene:—

UPPER.
- Upper fresh-water limestone marls and silicious millstone (*burrstone*).
- Upper marine-sands or Fontainebleau sandstone and sands.

MIDDLE { Lower fresh-water limestone and marl, or gypseous series.
Sandstone and sands, with marine sands (*sables moyens*).
Coarse sandy limestone, with marine shells (*calcaire grossier*).
Hard fresh-water limestone (*calcaire silicieux*).

LOWER { Lower sands, with marine shelly beds (*lits coquillier*).
Lower sands, with lignite and plastic clay (*argile plastique*).

137. As with the Paris and English deposits, so with the other tertiary basins of southern France, Spain, Austria, Hungary, Italy, &c.—all of them exhibiting an irregular succession of clays, sands, limestones, marls, gypsum, and lignites, which, when examined lithologically and palæontologically, are clearly referable to the same period of formation. Among the most remarkable features of foreign tertiaries are the *infusorial* and *nummulitic* strata—the former constituting such rocks as the "tripoli" of Bohemia and Virginia, and the latter the "nummulitic limestones," so abundant in southern Europe, Egypt, and Asia. The tripoli consists almost entirely of the silicious coverings of infusorial animalcules, and is often of great thickness, as at Richmond in Virginia, where it is nearly thirty feet; and the nummulitic limestone, which is composed of coin-shaped (*nummus*, a coin) foraminiferous shells, is perhaps the most important of ter-

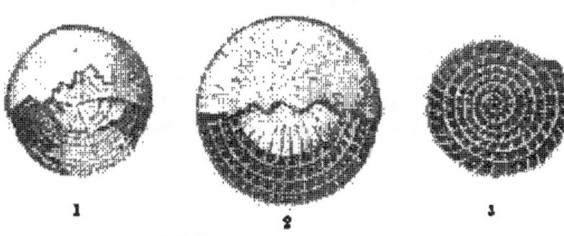

1, 2, Nummulites; 3, Section of do.

tiary strata. Respecting this limestone, which was till recently regarded as belonging to the cretaceous system rather than to the base of the eocene tertiaries, Sir Charles Lyell remarks that "it often attains a thickness of many thousand feet, and extends from the Alps to the Apennines. It is found in the Carpathians, and in full force in the north of Africa—as, for example, in Algeria and Morocco. It has also been traced from Egypt into Asia Minor, and across Persia by Bagdad to the mouths of the Indus. It occurs not only in Cutch, but in the mountain-ranges which separate Scinde from Cabul; and it has been followed eastward into India." From India its course has been recently traced onward to the Philippine Islands, thus demonstrating the existence of one of the most gigantic accumulations from the smallest of organic causes. A similar deposit of great extent occurs in North America, but in this case the organism is the *orbitolite* (globe-shaped), and hence it is there known as the orbitoidal limestone.

138. With respect to the extent and distribution of the lower tertiaries—laying aside the nummulitic limestone, which is in some respects a peculiar and unique development—we have as yet no certain knowledge. As there is often no perceptible mineral distinction between many clays, sands, and gravels, it is only by their imbedded fossils that geologists can determine their tertiary or post-tertiary character. Many accumulations at present regarded as superficial may be found hereafter to be of older date; and thus it becomes difficult to fix with certainty the geographical limits of the system. So far as Europe is concerned, tertiary deposits have received considerable attention, and their area has been found to be much more extensive than was at one time supposed. In general, the deposits occupy well-defined tracts or basins; hence the frequent reference in works on geology to the "London basin," "Hampshire basin," "Paris basin," "Vienna basin," and other tertiary tracts in Europe. As far as discovery has gone, there are few countries in Europe where tertiary strata have not been detected; and while we regard those of England, France, Austria, and Italy as typical, we must ever bear in mind that considerable modifications may require to be made, as the tertiaries of India and North and South America come to be more closely examined. One important fact must not be lost sight of in drawing any general conclusions from the distribution of tertiary deposits—viz., that as the fauna and flora of the period approach in character the fauna and flora of existing nature, and that as the plants and animals of Europe, India, Australia, and South America, &c., all differ widely from each other, so may we expect similar differences among the fossil remains of these distant regions. And this, as will afterwards be seen, is fully borne out—the tertiary mammals of South America resembling the sloths, armadilloes, ant-eaters, and alpacas of that continent; those of Australia its marsupial kangaroos and opossums; while those of the Old World have more immediate relationship to its elephants, rhinoceroses horses, deer, and oxen. And here it may be remarked that the student cannot too early direct his attention to the laws which regulate the distribution of life on the globe, and be able to distinguish clearly between *identity* and *representation* of species. During the Palæozoic and Mesozoic epochs there appears to have been a greater identity of species over wide areas; during the present period the areas are more circumscribed, and the species in one

region are only the representatives of those inhabiting another—that is, are specifically different in form, but discharge the same functions in the economy of nature. Thus the elephant of India is only represented by, and not identical with, the elephant of Africa; the lion and tiger of Asia are represented by the puma and jaguar of America; and the African ostrich finds its representative in the emu of Australia.

139. The igneous rocks associated with the system are, with the exception of a few doubtful cases, all of true volcanic origin. In England the tertiary strata have suffered no displacement or change from igneous action; but in central France, along the Rhine, in Switzerland, in Hungary, and in Italy, there are ample evidences of volcanic activity during the deposition of the system. The crateriform hills of Auvergne and the Rhine present the finest examples of this activity, and form as it were a connecting link between the secondary traps and the products of existing volcanoes. In their mineral composition the tertiary traps are chiefly trachytic —graduating from a compact felspathic greystone to a scoriaceous tufa, but in few instances presenting the dark, augitic, and basaltiform structure of the carboniferous traps, or the amygdaloidal and porphyritic texture of those associated with the old red sandstone and silurian.

140. As already stated, the organic remains of the system are all of *cainozoic* types—that is, either closely resemble, or are identical with, existing genera and species. Of course, since the commencement of the Eocene period, many forms of life have died away, and it is to these extinct families, rather than to those still surviving, that we shall now direct attention. The Flora of the tertiary exhibits few marine species—the nature of the deposits being apparently unfavourable for their preservation; but the fluvio-marine beds contain remains that can be referred to the cycads, palms, conifers, leguminosæ, willows, elms, sycamores, &c. Detached leaves, fruits, seeds, and seed-vessels are common in the London basin; and the lignites of the Continent exhibit the true dicotyledonous structure. Such names as *carpolithes* (*carpos*, fruit), *cupressinites* (*cupressus*, the cypress-tree), *nipadites* (like the nipa of Bengal), *faboidea* (*faba*, a bean), *leguminosites* (*legumen*, a pod), *tricarpellites*, *chara*, and the like, sufficiently indicate the external appearance and supposed alliances of these vegetable fossils. In the lignitic or tertiary coal-beds of America, New Zealand, and other regions, the plant-remains are generally closely allied to or identical with the genera still growing there; though in most lignites there is always a certain proportion of the plants belonging to extinct species. Of

ORGANIC REMAINS.

the Fauna, the invertebrate orders—infusoria, foraminifera, corals, star-fishes, sea-urchins, serpulæ, barnacles, crustacea, and shell-

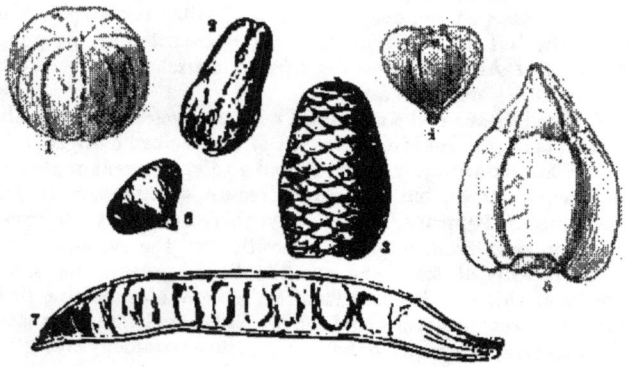

1. Cucumites; 2. Faboidea; 3. Petrophiloides; 4. Cupanoides; 5. Nipadites;
6. Leguminosites; 7. Micheliæ.

fishes—are extremely abundant, both numerically and in point of species. So closely related are many of the testacea to those of our present seas, that, as formerly stated, the groups, *eocene*, *miocene*, &c., have been instituted on the percentage of existing shells found in their strata. Thus:—

Pleistocene,	from	90 to 98	living species.	
Pliocene,	,,	60 to 80	,,	,,
Miocene,	,,	20 to 30	,,	,,
Eocene,	,,	1 to 3	,,	,,

With respect to the *fishes* of the tertiary epoch, "they are so nearly related," says M. Agassiz, "to existing forms, that it is often difficult, considering the enormous number (above 8000) of living species, and the imperfect state of preservation of the fossils, to determine exactly their specific relations. In general, I may say that I have not yet found a single species which was perfectly identical with any marine existing fish, except the little species (*mallotus villosus*) which is found in nodules of clay, of unknown geological age, in Greenland." The most common *ichthyolites* in the English tertiaries are the shark-like teeth of gigantic placoids which seem to have thronged these waters. Among the *reptiles*, the most abundant are fresh-water and marine turtles (*chelonia* and *platemys*), with true analogues of the existing crocodile and gavial. Of *birds*, several species have been described, chiefly from the Paris tertiaries. Of these, the eocene conglomerates of Meudon have recently yielded remains of a gigantic bird (*gastornis Parisiensis*), apparently intermediate between the swimmers and runners, the leg-bone indicating a bulk fully equal to that of the ostrich ; and in the miocene strata have been found several others that would seem to be connected with the genera buzzard, quail, curlew, sea-lark, kingfisher (*halcyornis*), pelican, vulture (*lithornis*), and condor (*pelagornis*); while many undetermined fragments of bird-bones are merely as yet ranked under the general designation of *ornitholites*. Of the *mammalia*, every existing order has had its tertiary representatives, though many of these genera are now extinct. Thus, the *marsupialia* (pouch-nursing), by several species allied to the kangaroo and opossum; the *cetacea* (whales) are represented by several species; the *edentata* (toothless animals), by gigantic analogues of the sloth, armadillo, and

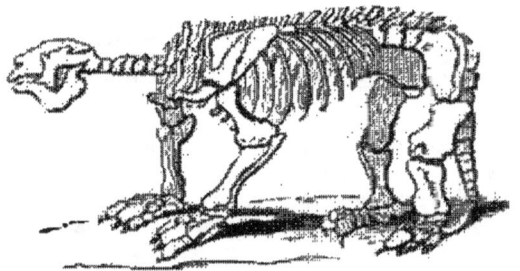

Megatherium.—Upper Tertiaries of South America.

ant-eater, as *megatherium, megalonyx*, &c.; the *ruminantia* (cud-chewers), by several species of elk, stag, antelope, buffalo, ox, &c.;

Mastodon.—Upper Tertiaries of Northern Hemisphere.

the *pachydermata* (thick-skins), by numerous forms—as deinotherium, mastodon, elephant, hippopotamus, rhinoceros, hog, tapir, and a number of creatures resembling tapir—as *palæotherium* (*palaios*, ancient, and *therion*, wild beast), *anoplotherium*, &c.; the *rodentia* (gnawers), by a number of species allied to the

Recovered Outlines—Ziphodon, Anoplotherium, Palæotherium.—Eocene Tertiaries of Paris Basin.

beaver, hare, rat, squirrel, &c.; the *carnivora* (flesh-devourers), by species of bear, hyæna, fox, dog, seal, &c.; the *insectivora* (insect-eaters), by remains of a species of mole; the *cheiroptera* (hand-winged), by two or three species of bat from the gypsum beds of Montmartre; and the *quadrumana* (four-handed), by one or two instances from the eocene beds of England apparently related to the Old World monkeys. Thus, every order of mammal, with the exception of man, has its representative during the tertiary epoch—differing it may be in certain species, but still presenting, on the whole, such a facies of resemblance, that one feels he is approaching the confines of existing nature.

141. Respecting the distribution of sea and land, and the climatal conditions of the world during the deposition of the tertiary strata, it is difficult to arrive at any satisfactory conclusion. It is certain, however, that one or other of the groups is to be found in every region; that in some instances they are strictly marine, and in others as decidedly fresh-water; while in many basins, as in England and France, they are partly freshwater and partly marine (*fluvio-marine*), as if there had been frequent submergences and elevations, or, at all events, periods when fresh-water inundations prevailed in the areas of deposit. As to physical conditions, the cycads and palms and monkeys of the London basin give evidence of a genial climate—a fact further corroborated by the huge pachyderms of the Paris strata,

the gigantic edentata of South America, and larger marsupials of Australia. On the whole, there seem to have been over a vast zone of Europe and Asia wide areas of shallow seas thronged with the humbler forms of marine life, and low sunny islands crowned with cycads and palms; broad estuaries prowled in by sharks; rivers swarming with crocodiles; open pasture-plains for the horse and buffalo; and dense woody jungles for the mastodon, deinotherium, and rhinoceros.

142. The industrial products of the system are building-stone and marbles of various quality; pipe and potter's clay in abundance; gypsum, or the well-known "plaster of Paris;" and the highly-valued *burr* millstone of France, which is obtained from the upper fresh-water series of the Paris basin. Lignite or "brown coal" is also worked in many tertiary districts; and amber, which is a fossil gum or gum-resin, is likewise obtained from the lignitic beds of the system.

Pleistocene Group.

143. This group, as the name implies, is intended to embrace all tertiary accumulations, the organic remains of which are chiefly referable to existing species. In the present state of geological knowledge, it is impossible to define with precision the limits of pleistocene tertiaries, and all that can be attempted is to arrange under one head the clays, sands, gravels, and boulders generally known as the "drift formation." As a whole, there is no class of rocks so perplexing, or whose origin is involved in greater obscurity, than this "drift" or "boulder-clay" — the "diluvium" of the earlier geologists. Composed in some districts of irregular ridges and mounds of sharp gravelly sand; in others of expanses of pebbly shingle; and more generally, perhaps, of various coloured clays, enclosing, without regard to arrangement, water-worn blocks or *boulders* of all sizes, from a pound to many tons in weight,—it is evident that it does not owe its origin to the ordinary sedimentary operations of water. It is also for the most part unfossiliferous; marine shells being found, and that very sparingly, only in certain sands and clays belonging apparently to the earlier and later portions of the epoch. Under these circumstances it will be sufficient for the purposes of the beginner to describe the leading phenomena connected with its occurrence in the British Islands and north of Europe, and to direct his attention to some of the theories that have been advanced to account for its formation.

144. It has been already stated that the pleistocene group consists of accumulations of clays, sands, gravels, and boulder-stones —the latter sometimes lying detached or in masses, but more frequently enclosed in the clays without regard to gravity or any other law of arrangement. We say "accumulations of clays, sands, &c.," because these seldom or never appear in regular strata, but here in masses, and there spread over wide tracts, as if brought together by some unusual and extraordinary operation of water. These unusual appearances have long and largely engaged the attention of observers; hence the variety of designations, such as "diluvium," "diluvial drift," "northern drift," "erratic-block group," and "boulder formation." When we examine the group as it occurs in Britain, we find it in some tracts (eastern counties of England) an open gravelly drift, consisting of fragments of all the older rocks, from the granite to the chalk inclusive. In other districts, as the middle counties of Scotland, large areas are covered with a thick, dark, tenacious clay, locally known by the name of "till," and enclosing rounded and water-worn boulders, as well as angular fragments of all the older and harder rocks—granite, gneiss, greenstone, basalt, limestone, and the more compact sandstones. The boulders are of all sizes, are most frequently rounded and water-worn, and are distributed throughout the mass without any regard to sedimentary deposition. In other localities, both in England and Scotland, we find large areas covered by loose rubbly shingle and sand; the shingle and sand often appearing in mound-like ridges, or in flat-topped irregular mounds (the *kaimes* of Scotland, the *eskars* of Ireland, and *ösars* of Sweden), as if the originally gravel deposit had been subsequently furrowed and worn away by currents of water. Occasionally districts are thickly strewn with boulders which rest on the bare rock-formations, without any accompanying clays or sands; and at times only a single gigantic boulder will be found reposing on some height, as sole evidence of the drift formation. When we come to examine the clays and sands more minutely, we find them partaking less or more of the mineral character of their respective districts. Thus, the boulder-clays of our coal districts, though thickly studded with boulders of distant and primitive origin, are usually dark-coloured, and contain fragments of coal, shale, and other carboniferous rocks. The same may be remarked of old and new red sandstone areas, where the clays are usually red; and of oolitic and chalk tracts, where they assume a yellowish or greyish aspect.

145. In addition to what has been stated respecting the com-

position of the drift, it may be remarked that the sands seldom exhibit lines of stratification, and that the clays are rarely or never laminated. Occasionally sands and clays alternate, or a dark-coloured clay may be overlaid by a lighter-coloured one; but more frequently sands and clays occur *en masse*, enclosing curious "nests" or patches of gravel, and crowded accumulations of boulder-stones. On examining the surfaces of many of these boulders we find scratches and groovings, as if they had been rubbed forcibly over each other in one direction; and what is still more curious, the surfaces of the rocks on which the boulder-clay reposes are all less or more smoothed, and marked with bold linear scratches and furrows, as if the boulders had been forcibly carried forward, and had scratched and grooved them during the passage. Again, these scratchings and groovings generally trend in lines parallel to the hill-ranges and valleys in which they occur, or radiate in all directions if there be some high and commanding eminence in a district which has served as a centre of dispersion. Moreover, most of the hills, as in Britain, present a bare bold craggy face to the west and north-west, as if worn and denuded by water, while their slopes to the east and south-east are usually masked with thick accumulations of clay, sand, and gravel. This appearance, generally known by the name of "crag and tail," is ascribed to the same moving forces which transported the enormous boulders of the Drift, and furrowed the surfaces of the rock-formations over which they were borne.

140. Taking all these phenomena into account, it is quite clear that pleistocene accumulations owe their origin to no ordinary operations of water. We can conceive of no current sufficiently powerful to transport boulder blocks of many tons in weight over hill and dale for hundreds of miles; of no sedimentary condition that would permit boulders and clays to be huddled up in the same indiscriminate mass; while the smoothing and grooving of rock-surfaces point to long-continued action, and not to any violent cataclysm, even could we conceive of one sufficiently powerful to transport the blocks and boulders. There is only one set of physical conditions with which we are acquainted sufficient to account for all the phenomena — namely, an arctic or frozen climate, with glaciers to wear and furrow the surface of the land, and with icebergs and ice-floes dropping off shore to transport the eroded material into deeper water; and it is now to such conditions that geologists turn for a solution of the boulder formation. After the deposition of the Pliocene, it would seem that the latitudes of Britain and the north of

DRIFT FORMATION.

Europe underwent a vast revolution as to climate, and that some new arrangement of sea and land took place at the same period. At all events, the large mammalia of the earlier tertiaries disappeared, and the land was submerged to the extent of several thousand feet, for we now find water-worn boulders on the tops of our highest hills. A cold and boreal climate accompanied this submergence—an ice-mantle like that of Greenland spread over the land, wearing and smoothing and grooving, and icebergs broke off shore, bearing their burdens of eroded material and boulders to be dropped in the submerged areas. How long this process continued it is impossible to determine; but after a very long period a gradual elevation of the submerged lands took place; our hill-tops and ranges appeared as islands, and our valleys as firths and straits. These islands were still covered with ice, and during a brief summer avalanches descended, glaciers smoothed the hill-sides, and left the debris as *moraines* of sand and gravel; while the icebergs and ice-floes ground their way through the firths, disturbing, adding to, and in some measure re-arranging, the debris that had been dropped during the period of submergence. As the ele-

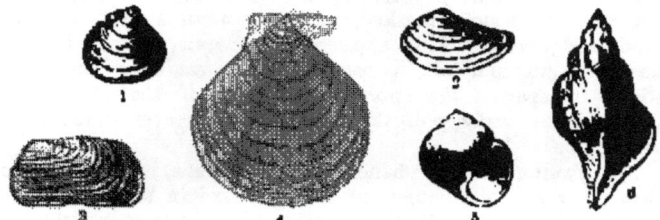

Boreal shells in the Upper Drift of the Clyde—Smith
1. Astarte borealis; 2. Leda oblonga; 3. Sanicava rugosa; 4. Pecten Islandicus;
5. Natica clausa; 6. Trophon clathratum

vation continued, new surfaces were exposed, the western fronts of our hills were wasted by waves and swept bare by currents, and the soft material of the sea-bottom, as it rose above the waters, was washed away and carried to areas of sea-bottom not yet elevated above the waters. We say the "western," or rather "north-western," front of our hills; for, taking the phenomena of crag and tail into account, the direction of the groovings on rock-surfaces, and other kindred appearances, it is evident that in Britain the transporting currents passed from north and west to south and east. It is thus that we find granitic and gneiss boulders from the Scottish Highlands now spread over the plains of Fife and Mid-Lothian, and blocks from the hills of Cumberland scattered over the moors of Yorkshire. In the north of Europe the drift has taken a more southerly course, and thus boulders from Lapland and Finland are spread over the plains of Russia and Poland; and granites from Norway now repose on the flats of Denmark and Holstein. Occasionally, as in Switzerland, the drift appears to radiate from a centre; and this we can readily conceive as the Alps rose isolated in a glacial sea, and annually dispersed their glaciers and icebergs in every direction.

147. In process of time the land was elevated to its present level, another distribution of sea and land took place, and the glacial epoch passed away. A new flora and fauna suitable to those new conditions were then established in Europe; and these, with the exception of a few that have since become extinct, are the species which now adorn our forests and people our fields. Hitherto we have spoken only of the "Drift" as exhibited in northern Europe; but similar phenomena are manifested in Canada and the northern States of America. Again, when we turn to the Antarctic Ocean, analogous appearances present themselves in Tierra del Fuego and Patagonia; thus showing that, as at the present day glaciers wear and waste the land-surface of boreal regions, and icebergs float towards warmer latitudes, and drop their burdens of sand, mud, and boulders on the sea-bottom; so during the pleistocene epoch the same agencies were at work discharging the same functions, and producing analogous results.

RECAPITULATION.

The tertiary system, as described in the preceding chapter, embraces all the regular strata and sedimentary accumulations which lie between the Chalk and the close of the Boulder or Drift Formation. Its organic remains are all of recent or *Cainozoic* types, and it has been subdivided into four groups, according to the numerical amount of existing species found imbedded in its strata, thus—

        Pleistocene—remains, mostly of existing species.
        Pliocene—remains, a majority of existing species.
        Miocene—remains, a minority of existing species.
        Eocene—remains, few (or the dawn) of existing species.

In their mineral composition and succession these groups present great variety—consisting of clays, sands, marls, calcareous grits,

RECAPITULATION.   163

limestones, gypsum, and beds of lignite, with evidences of frequent alternations from marine to fresh-water conditions. On the whole clays and limestones prevail, and many of the latter are of very peculiar character, as the fresh-water burrstones of Paris, the gypsum or sulphate-of-lime beds of Montmartre, the infusorial tripoli of Bohemia, and the nummulitic limestone of the Alps, Egypt, and India. Separating the older or true tertiaries from the pleistocene or boulder group, it may be said that the former are found less or more in almost every country, though often confined to limited areas, as if originally deposited in inland seas or estuaries. These well-defined deposits are usually termed "basins;" hence the frequent allusion to the London and Paris basins, in which there are frequent alternations of marine and fresh-water beds, as if at certain stages fresh-water inundations had prevailed in the areas of deposit. The tertiaries of England, France, Switzerland, and Italy, are those that have been most fully investigated, and though differing in the composition and succession of their strata, are generally regarded as finding their equivalents in those of England, which may be briefly grouped as under:—

PLEISTOCENE. { Fossiliferous clays and sands of Clyde, Forth, &c.
{ Boulder or drift formation.

PLIOCENE.   Mammaliferous and Red crag.

MIOCENE.   Coralline crag.

EOCENE.   { Fluvio-marine beds of Isle of Wight.
             { Bagshot sands.
             { London clay.
             { Bognor beds.
             { Plastic clay.

As already stated, the organic remains of the system belong in greater part to existing species, and thus among the *plants* we find the leaves, fruits, and seed-vessels of palms, cycads, pines, and, for the first time, of true exogenous timber-trees; while among the *animals* we discover species of every existing order, invertebrate and vertebrate, with the exception of man. The most characteristic feature of the fauna is perhaps the abundance of gigantic quadrupeds—in European tertiaries, of mastodons, elephants, deinotheriums, palæotheriums, rhinoceroses, &c.; in South America, of megatheriums, megalonyxes, glyptodons, &c.; and in Australia, of animals allied to the marsupials of that continent, but of more gigantic proportions. The names given to those animals have reference in general to some striking peculiarity of structure, size, or appearance ; as *mastodon*, from the pap-like crowns of its teeth (Gr. *mastos*, a nipple ; *odous*, a tooth) ; *glyptodon* (*glyptos*, carved or sculptured), from the curious markings of its teeth ; *megalonyx* (*megale*, great, and *onyx*, a claw) from its large claws ; *deinotherium* (*deinos*, terrible), terrible wild beast ; *diprotodon*,

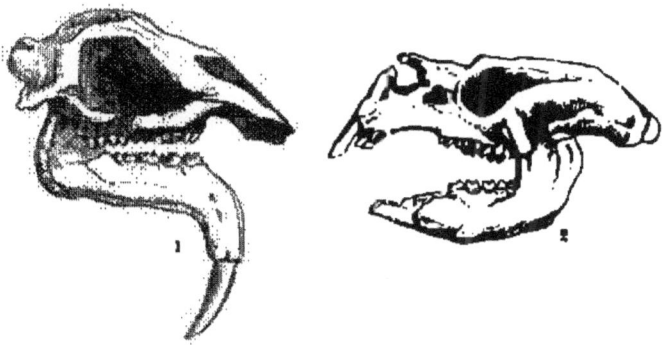

1. Deinotherium giganteum ; 2. Diprotodon Australis.

two front teeth ; *megatherium*, huge wild beast ; and so on of others. In respect of its fossils, the tertiary era presents a remarkable difference compared with those of the chalk, oolite, or coal. During these epochs the plants and animals in every region of the globe presented a greater degree of sameness or identity; whereas, during the tertiary epoch, geographical distinctions and separations like those now existing began to prevail ; hence the difference between the tertiary mammals of Europe and those of South America, which represent its present sloths, ant-eaters, and armadilloes. Whatever the conditions of other regions during the deposition of the tertiary strata, we have evidence from the palms, cycads, huge pachyderms, and monkeys, that in the latitudes now occupied by England and France a greatly more genial climate prevailed ; and that at the close of the pliocene strata, these conditions were followed by those of a boreal or glacial character, which gave rise to the boulder or drift formation. As a separate group, the pleistocene, in its unfossiliferous clays, its huge water-worn boulders, its smoothed and scratched rock-surfaces, and other kindred phenomena, gives evidence of a long period when these latitudes were subjected to arctic conditions, when glaciers covered its hills and islands, smoothing and grooving the rock-surfaces on which they rested, and icebergs floated over its waters, or ground their way through its firths and straits, dropping, as they melted away, their burdens of clay, sand, and boulders on the deeper sea-bottom.

XVI.

POST-TERTIARY SYSTEM, COMPRISING ALL ALLUVIAL DEPOSITS, PEAT-MOSSES, CORAL-REEFS, RAISED BEACHES, AND OTHER RECENT ACCUMULATIONS.

148. HAVING treated the Boulder-drift as the latest member of the Tertiary system, we now proceed to describe, under the term *Post-Tertiary* or *Quaternary* (Lat. *quatuor*, four), all accumulations and deposits formed since the close of that epoch. However difficult it may be to account for the conditions that gave rise to the "Drift," there can be no doubt regarding the agencies which have been at work ever since in silting up lakes and estuaries, forming peat-mosses and coral-reefs, laying down beaches of sand and gravel, throwing up volcanic hills and islands, slowly submerging some tracts of land, and as gradually elevating others above the waters of the ocean. At the close of the Pleistocene period, the present distribution of sea and land seems to have been established; the continents presenting the same surface-configuration, and the ocean the same coast-line, with the exception of such modifications as have since been produced by the atmospheric, aqueous, and other causes described in Chapter II. At the close of that period the earth also appears to have been peopled by its present flora and fauna, the only change being some local removals of certain plants and animals, and the general extinction of a few species, whose remains are found imbedded in a partially petrified or *sub-fossil* state in post-tertiary accumulations. We are thus introduced to the existing order of things; and though our observations may extend over a period of many thousand years, yet every phenomenon is fresh and recent compared with those of the epochs already described. With the exception of volcanic lavas, deposits from calcareous and silicious springs, some consolidated sands and old coral-reefs, we have now no hard strata—the generality of post-tertiary accumu-

lations being clays, silts, sands, gravels, and peat-mosses. As they are scattered indiscriminately over the surface, it is impossible to treat them in anything like order of superposition; hence the most intelligible mode of presenting them to the beginner, is to arrange them according to their composition, and the causes obviously concerned in their production. Adopting this plan, the principal agencies and their results may be classed as follows:—

FLUVIATILE.
- Accumulations of sand, gravel, and alluvial silt in valleys and along river-courses.
- Terraces of gravel, &c., in river-valleys, marking former water-levels.
- Deposits of sand, silt, &c., in estuaries, forming "deltas."

LACUSTRINE.
- Lacustrine accumulations now in progress.
- Lacustrine or lake silt filling up ancient lakes.
- Shell and clay marl formed in ancient lake-basins.

MARINE.
- Submarine (deep-sea) deposits and accumulations.
- Marine (littoral) silt, sand-drift, shingle beaches, &c.
- Raised or ancient beaches; submarine forests.

CHEMICAL.
- Calcareous deposits, as calc-tuff, travertine, &c.
- Silicious deposits, as silicious sinter, &c.
- Saline and sulphurous deposits from hot springs, volcanoes, &c.
- Bituminous exudations, as pitch-lakes and the like.

ORGANIC.
- Vegetable—peat-mosses, jungle-growth, vegetable drift.
- Animal—shell-beds, coral-reefs, osseous breccia, &c.
- Soils—admixtures of vegetable and animal matters.

IGNEOUS.
- Elevations and depressions caused by earthquakes.
- Displacements produced by volcanic eruptions.
- Discharges of lava, scoriæ, dust, and other matters.

Carefully reviewing the above synopsis, and bearing in mind what was stated in Chapter II. respecting the causes now modifying the crust of the globe, the student need be presented with little more than a mere indication of these accumulations.

Fluviatile Accumulations.

149. Under this head (*fluvius*, a river) are comprehended all accumulations and deposits resulting from the operations of rivers. We have already seen (pars. 17 and 18) how streams and rivers cut for themselves channels, glens, and valleys, and transport the eroded materials in the state of mud, sand, and gravel to some lower level. During inundations and freshets, some of this debris is spread over the river-plains: in ordinary cases, some of it is deposited in lakes and marshes, should such

lie in its course; and in all cases a notable proportion is lodged in estuaries or carried out into the ocean. The natural tendency of rivers being thus to deepen their channels, and spread the eroded material over the lower levels, all river-valleys will in course of time become dry plains, even though originally consisting of marshes and chains of lakes. Such operations have been going on since the land received its present configuration; and thus we have fluviatile deposits of vast antiquity, as well as accumulations whose origin is but of yesterday. Such alluvial tracts as the "carses" and "straths" of Scotland, and the "dales" and "holmes" of England, have been formed partly in this way, and partly by upheaval of the deltas and estuaries into which their rivers are discharged; and in many of these have been found the bones of elephants, rhinoceroses, wild boars, deer, wild oxen, bears, wolves, beavers, and other animals long since extinct in the British Islands, as well as the remains of seals, whales, and other marine creatures which only occasionally frequent our shores. Such accumulations are often of great thickness, and consist for the most part of alluvial silt, masses of gravel and shingle, with occasional beds of fine dark-blue unctuous clay, and layers of peat-moss and shell-marl.

150. In most of the inland valleys of this and other countries there appear, belting their slopes, long level terraces, composed of sand, shingle, and silt. Such terraces give evidence of former water-levels, and point to a time when the valley was occupied by a lake at that height, or when the plain stood at that level, and before the river had worn its channel down to the present

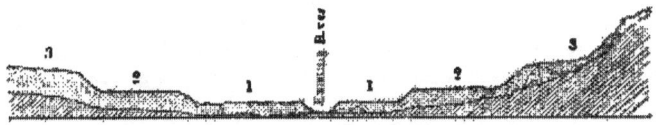

River or Valley Terraces, marking former water-levels.

depth. River-terraces must not be confounded with the raised beaches which fringe many parts of our coasts and estuaries; for, though both are in one sense ancient water-levels, the former may be local and partial, while the latter are general and uniform. Besides, the remains found in the one are of terrestrial or fresh-water origin; in the other they are strictly marine. These terraces have long attracted attention, and point to a time when many of our fertile valleys were chains of lakes and morasses, which have been drained and converted into alluvial land by the natural deepening of the river-channels.

151. At the mouths or in the estuaries of all existing rivers there have been accumulating, since sea and land received their present configuration, deposits of mud, sand, gravel, and vegetable debris. In course of time these deposits constitute large expanses of low alluvial land, known as "deltas," the most notable instances of which are those of the Rhine and Po in Europe, of the Nile and Niger in Africa, of the Ganges and Chinese rivers in Asia, and of the Mississippi and Amazon in America. Many of these deposits are of vast extent, and, with the exception of what is taking place at the bottom of the ocean (of which we know almost nothing), they are of all modern formations the most important in modifying the crust of the globe. Where a river discharges itself into a non-tidal sea, like the Po into the Gulf of Venice, the delta will be mainly of fluviatile origin; but where the discharge is into a tidal sea, like the Ganges into the Bay of Bengal, the deposit will be partly fluviatile and partly marine (*fluvio-marine*). Further, the deltas of tropical rivers subject to periodical inundations are, during the dry season, low flat tracts full of swamps, creeks, and mud-islands, which nourish the rankest jungle-growth, herds of gigantic amphibia, shell-beds, and shoals of fishes. On the return of the wet season, many of these plants and animals are buried where they grew, or are swept forward into the ocean. We have thus a complex set of agents—rivers, tides, waves—the drift from inland, the drift from the sea, and the growths of plants and animals *in situ*. All these conjoined render estuarine deposits extremely perplexing and irregular in their composition; and though in general terms they may be said to consist of mud, clay, sand, gravel, and vegetable debris, intermingled with organisms of terrestrial, fresh-water, and marine origin, yet scarcely two of them present one feature in common. In their fossil contents they must also vary, partly according to the latitude in which they are situated, and partly according to the nature of the country through which their rivers flow—the Ganges, for example, entombing in its delta the palms, tree-ferns, elephants, tigers, and lions of India; the Niger, the palms, the elephant, hippopotamus, rhinoceros, giraffe, camel, and ostrich of Africa; while the Mississippi floats down the pines, buffaloes, elk, deer, and beavers of North America.

Lacustrine or Lake Deposits.

152. Lacustrine deposits (*lacus*, a lake) are those found either

in existing lakes, or occupying the sites of lakes now filled up. Lakes are found in every region of the world, and act as settling-pools or filters for the rivers that flow through them. A river on entering a lake may be turbid and muddy, while the water that flows from it is limpid and clear as crystal. The mud or sand settles down as silt, and successive depositions of silt with inter-mixtures of vegetable drift and peat-moss and marl, constitute the ordinary composition of lacustrine accumulations. Situated in plains or valleys, a lake serves in general as a basin of reception to several streams and rivers. The mud borne down by these streams settles at their mouths and forms small deltas, which in process of time are covered with rushes, reeds, sedges, and other marsh plants; new accumulations of sediment push their way into the centre of the lake, and new growths of marsh plants arise. The annual growth and decay of these plants form beds of peat; while fresh-water shells, microscopic organisms, and calcareous springs, combine to elaborate layers of marl. These agencies, acting incessantly, are gradually shoaling and silting up all lakes; lessening the areas of some, converting others into marshes, and these again into alluvial meadow-land. Silted-up lakes are rife in every country, and a great proportion of our alluvial valleys are but the sites of marshes and lakes filled up and obliterated by the process above described. The organic remains found in lake-deposits are strictly fresh-water and terrestrial—fresh-water shells, as the limnæa, planorbis, and paludina, in the marls; marsh plants, as the reed, bulrush, and equisetum, in the peat-moss; drift, or terrestrial plants, as the birch, alder, hazel, oak, pine, &c., in the silt; with bones and sometimes complete skeletons of the Irish deer, red-deer, wild-ox, horse, bear, beaver, otter, and other mammalia. In many of the lake-deposits of Britain, Ireland, France, Belgium, and Switzerland, tree-canoes, stone battle-axes, bone weapons, and other objects of human art, have been discovered, all pointing to the recent period of geology, though chronologically of vast antiquity, and far beyond the written records of our race.

Marine Deposits.

152. The marine deposits of the modern epoch naturally divide themselves into three great classes—those taking place under the waters of the ocean, as deep-sea muds, sandbanks, and shoals; those collecting along the sea-margin, as sand-drift and shingle-

beaches; and those, like ancient beaches, now elevated above the existing sea-level. Respecting the first class of deposits, we know very little indeed, few parts of the ocean having been sounded for geological purposes; and even where sounded, the indications are too obscure to warrant any definite conclusion. So far as dredgings and soundings enable us to decide what is going on under the waters, submarine deposits appear to be extremely varied: here soft slimy mud, there light-coloured clay, with shells; here shelly sand, replete with minute foraminifera and broken corals—there sandy shoals and gravel-banks; and over the whole, elevations and depressions as irregular and varied as those of the dry land. Such irregularities of sea-bottom, conjoined with the configuration of sea and land, give rise to numerous currents, and these currents not only distribute the submarine debris, but transport the products of one region to another. The principal ocean-currents are the tides, with all their varied ramifications; the Equatorial currents, with their westward flow; the Gulf Stream, setting out from the Gulf of Mexico north-eastward across the Atlantic; the Black Stream of the Pacific, setting northward to the shores of Japan; and the Polar currents, which set in from either pole towards the equator. The tidal currents are perpetually shifting and redistributing the deposits along the sea-bottom; the Gulf Stream is as regularly transporting tropical products to temperate regions; and the polar currents carry with them icebergs and ice-floes laden with rocks and gravel, which are dropped on the sea-bottom as the ice melts away in warmer latitudes. All these agents are incessantly at work; and thus deposits are now accumulating along the bottom of the ocean, which, if raised into dry land, would equal in extent any of the older formations.

154. Marine silt, sand-drift, shingle-beaches, and the like, are the terms usually applied to accumulations which have taken place, or are still in progress, along the present shores of the ocean. Waves and tidal transports are the agents to which these owe their origin; they occur in bays and sheltered recesses, and, as strictly *marine* formations, are not to be confounded with the silt of estuaries and river embouchures. Around the shores of our own island, and, in fact, along the shores of every other country, the tides and waves are wasting away the land in some localities, and transporting the debris to sheltered bays and creeks, there to be laid down as mud-silt, sand, or gravel. This process must have been going forward since sea and land acquired their present distribution, and thus many of these accumulations are both of vast extent and great antiquity. As examples of marine

silt, we may point to the "warp" of the Humber, to the fens and marshes of Lincolnshire and Cambridge, and to the low plains of Holland and Denmark, which are all the immediate formation of the German Sea. Of sand-drift, which is first accumulated by the tides and waves, and subsequently blown into irregular heights and hollows by the winds, we have instructive instances in the Bay of Biscay near the Garonne, between Donegal Bay and Sligo Bay in Ireland, and between the Tay and Eden in Scotland. Of these recently-formed silts and sand-drifts, thousands of acres lie waste and worthless; but of the older portions large tracts have been reclaimed, and are now under the plough of the farmer. Their organic remains are of all ages, from the drift of the last tide to the half-petrified bones of whales and shells of mollusca not now to be found in these seas.

155. All along the shores of the British Islands, as well as along the shores of every other sea, there exists a level margin, more or less covered with sand and gravel. This constitutes the existing *beach*, or sea-margin; but above it, at various heights, are found, following the bays and recesses of the land, several similar margins or terraces known as "ancient or raised beaches." These give evidence of either elevation of the land or depression of the ocean, and point to times when sea and land stood at these successive levels. We have several notable examples along our own coast, at heights about nine, twenty-five, forty, sixty-three, one hundred and twenty feet, and even still higher elevations above the present sea-level; similar indications are found along the coast of Norway, the shores of the Baltic, the coasts of Siberia and Greenland, in the Bay of Biscay, and along the coasts of Spain; while elevated water-levels have also been discovered on the coasts of both Americas. In some districts these terraces are covered with sand, shells, and shingle; in other localities, a mere shelf or line along a hill-side bears evidence of the former presence of the tides and waves. As raised beaches point to successive elevations of the land, so do submarine forests give evidence of occasional depressions. We say *occasional* depressions, for as yet we have only a single line of these so-called forests, which occur at distant intervals along the Firths of Forth, Eden, and Tay, along some of the bays of Shetland, between the Tyne and Wear, at Hartlepool near the mouth of the Tees, at Hull on the Humber, and on the coasts of Hampshire, Devonshire, and Lancashire. In general, these submarine forests consist of a bed of peat or semi-lignite, from two to six feet in thickness, abounding in roots and trunks of trees in the lower portion, and in mosses and aquatic plants in the upper and lighter-coloured

portion. The trees are chiefly oaks (often of great dimensions), Scotch firs, alders, birches, hazels, and willows; and throughout are imbedded hazel-nuts, seeds of various plants, and the wing-cases of insects. The forests rest for the most part on dark-blue unctuous clay, and are overlaid by from twelve to twenty feet of marine silts and sands—thus showing, *first*, that the forest-growth had been formed at a higher elevation than the present seaboard; *secondly*, that after its formation and consolidation into peat, it had been submerged and overlaid by the sea-silts and sands; and *thirdly*, that after being covered by these silts, it had been re-elevated to its existing level. This submarine forest-growth also implies not only a greater extension of the British Islands than at present, but, from the great size of its trees and the nature of the imbedded insects, the existence of a somewhat warmer climate between the present day and the boulder-clay epoch.

Chemical Deposits.

156. Under this head we class all deposits arising from calcareous and silicious springs, all saline incrustations and precipitates, and all bituminous or asphaltic exudations. The most frequent deposits of a calcareous nature are calc-tuff and calc-sinter, stalagmites and stalactites, and travertine. *Calc-tuff*, as the name implies, is an open, porous, and somewhat earthy deposition of carbonate of lime from calcareous springs, and is found in considerable masses, enclosing fragments of plants, bones, land-shells, and other organisms. *Calc-sinter*, from the German word *sintern*, to drop, is of similar origin, but more compact and crystalline, and has a concretionary structure, owing to the successive films which are drop by drop added to the mass. *Stalagmites* and *stalactites*, already noticed in par. 23, are often of considerable magnitude in limestone caverns, and are here noticed as frequently enclosing the bones and skeletons of animals found in these caverns. *Travertine* (a corruption of the word Tiburtinus) is another calcareous incrustation, deposited by water holding carbonate of lime in solution. It is abundantly formed by the river Anio at Tibur, near Rome; at San Vignone, in Tuscany, and in other parts of Italy. It collects with great rapidity, and becomes sufficiently hard in course of a few years to form a light durable building-stone. As with deposits from calcareous, so with deposits from silicious springs—these forming silicious tufa and sinter in considerable masses, as at the hot springs or *Geysers* of Iceland (where it fills

fissures 12 and 14 feet in width), the Azores, and other volcanic regions.

157. In hot countries, incrustations of common salt, nitrate of soda and potash, and other saline compounds, are formed during the dry season in the basins of evaporated lakes, in deserted river-courses, upraised or ancient sea-levels, and in shallow creeks of existing seas. These incrustations go on from year to year, and in course of time acquire considerable thickness, or are overlaid by sedimentary matter, and there exhibit alternations like the older formations. Such deposits are common in the sandy tracts of Africa, in the river-plains and old sea-reaches on the upheaved coasts of South America, along the coasts of India, and in the salt lakes of Central Asia. With respect to springs and exudations of petroleum, asphalt, and the like, it may be remarked that they are too limited and scanty to produce any sensible effect on the bulk of the rocky crust, and are principally of geological importance, as throwing light on analogous products of earlier date.

Organic Accumulations.

158. Organic accumulations, as depending on the agencies described in para. 20 and 21, consist either of vegetable or of animal remains, or of an intimate admixture of both. The most important of those resulting from vegetable growth are peat-mosses, jungle-swamps, drift-rafts, and submerged forests. *Peat*, which is a product of cold or temperate regions, arises from the annual growth and decay of marsh plants—reeds, rushes, equisetums, grasses, sphagnums, and the like, being the chief contributors to the mass, which in process of time becomes crowned and augmented by the presence of heath and other shrubby vegetation. Peat-moss has a tendency to accumulate in all swamps and hollows; and wherever stagnant water prevails, there it increases, filling up lakes, choking up river-courses, entombing fallen forests, and spreading over every surface having moisture sufficient to cherish its growth. It occupies considerable areas in Scotland and England, though rapidly disappearing before drainage and the plough; but it still covers wide areas in Ireland. It is found largely in the Netherlands, in Russia and Finland, in Siberia, Canada, and British North America, and in insular positions, as Shetland, Orkney, and the Falkland Islands. It occurs in all stages of consolidation, from the loose fibrous turf of the present generation to the compact lignite-looking peat formed thousands of years ago. Besides the peculiar plants which consti-

tute the mass, peat-mosses contain trunks of oak, pines, alders, birches, hazels, and other trees, apparently the wrecks of forests entangled and destroyed by their growth, prostrated by storms, or felled by the hand of man. Bones and horns of the Irish deer, stag,

Megaceros Hibernicus, or Gigantic Irish Deer

ox, and other animals, are found in most of our British mosses, with occasional remains of human art, as tree-canoes, stone-axes, querns, and coins; and not unfrequently the skeleton of man himself. Some of these fossils are comparatively modern; others point to a period apparently coeval with the dawn of the human race. As with peat-mosses in temperate latitudes, so with the jungle-growth of tropical deltas, as those of the Niger, Ganges, and Amazon; so with the cypress-swamps (the "Great Dismal," for instance), of the United States; and so also with the pine-rafts and vegetable debris borne down by such rivers as the Mississippi, and entombed amid the silt of their estuaries. All are adding to the solid structure of the globe, and forming beds, small it may be in comparison, but still analogous to the lignites of the tertiary, and the coals of the oolitic and carboniferous eras.

159. Accumulations resulting from animal agency are universal and varied; but those of any appreciable magnitude are chiefly coral-reefs, shell-beds, and infusorial deposits. The nature and growth of the coral zoophyte has been already alluded to in par. 22, and we need here only observe the extent of its distribution in the Pacific, Indian, and Southern Oceans. Viewing a *coral-reef* as essentially composed of coral structure, with intermixtures of

drift-coral, shells, sand, and other marine debris, we find such masses studding the Pacific on both sides of the equator, to the thirtieth degree of latitude; abounding in the southern part of the Indian Ocean; trending for hundreds of miles along the northeast coast of Australia; and occurring in minor and scattered patches in the Persian, Arabian, Red, and Mediterranean Seas. In the Pacific, where volcanic agency is actively upheaving and submerging, coral-reefs are found forming low circular islands (*atolls*), fringing islands of igneous origin (*barrier-reefs*), crowning

Whitsunday Island, or Atoll.

others already upheaved (*coral ledges*), or stretching away in long surf-beaten ridges (the true *reef*) of many leagues in length, and from twenty to more than one hundred feet in thickness. Regarding them as mainly composed of coral, and knowing that the zoophytes can only add a few inches to the general structure during a century, many of these reefs must have been commenced before the dawn of the present epoch; and looking upon them as consisting essentially of carbonate of lime, we have calcareous accumulations rivalling in magnitude the limestones of the secondary formations. *Shell-beds*, like those formed by oysters, cockles, mussels, and other gregarious molluscs, are found in the seas and estuaries of every region, often spread over areas of considerable extent, and several feet in thickness. Dead shells, like the pearl-oyster of the Indian seas, for example, are also accumulated on certain coasts in vast quantities; and shell-sand, entirely composed of comminuted shells, is drifted for leagues along the shores of every existing sea. In fact, when we consider the myriads of testacea (Lat. *testa*, a shell) that throng the waters of the ocean, the rapidity with which they propagate their kind, and the indestructible nature of their shells, we are compelled to admit their accumulations to a place in the present epoch, as important as that which they held in any of the earlier eras. In treating of the

ORGANIC ACCUMULATIONS. 177

chalk and tertiary strata, we saw what an important part had been played in the formation of certain beds by infusorial organisms and minute foraminifera; and so far as the researches of microscopists have gone, it would appear that the same minute agencies are still at work in the silt of our lakes and estuaries, and in the shoals of our seas. What the eye regards as mere mud and clay, is found, under the lenses of the microscope, to consist of countless myriads of the silicious shields of diatoms, or the calcareous shells of foraminifera—a discovery whose limits will be further extended as the microscope becomes, as it soon must be, the inseparable companion of the geological inquirer.

160. Although coral-reefs, shell-beds, and infusorial deposits are the only accumulations of any magnitude arising from animal agency, yet what are usually termed *ossiferous gravels* (*os*, a bone, and *fero*, I yield), *ossiferous caverns*, and *osseous breccia*, are too important in a palæontological point of view to be passed over without some notice, however cursory. Sands and gravels containing masses of drift-bones, such as the tusks and grinders of the mammoth and elephant, the bones and teeth of the rhinoceros, hippopotamus, horse, bear, &c., and the horns and bones of the elk, stag, and wild-ox, are common in the valleys of Britain, in the river-plains of North America, and in the gravel cliffs of Siberia and the polar seas. Caverns occurring in the limestones of England, France, Belgium, Germany, Italy, North America, and Australia, are often replete with bones preserved in stalag-

Elephas primigenius, or Mammoth.

mitic incrustations, or in calcareous mud; and masses of drifted bones (osseous breccia) occur in rents and fissures, cemented together, and petrified by calcareous tufa. To this curious series of accumulations belong the mastodons of North America, the mammoths of Siberia, the dinornis or gigantic ostrich-like bird of

New Zealand, the elephants and elks of our own valleys, and the remarkable heterogeneous accumulations of elephant, hippopotamus, horse, bear, hyæna, deer, ox, stag, and other bones found in the limestone caverns of Yorkshire, Derbyshire, Somerset, and Devonshire. Most of these remains belong to animals now extinct in the countries where they occur, and point to a period immediately preceding, or even during the first commencement of the "Boulder-drift," and long anterior to the appearance of the human race. Occasionally, as in Belgium, France, and along the shores of the Mediterranean, human bones, the embers of fires, stone implements, and other traces of savage life, are

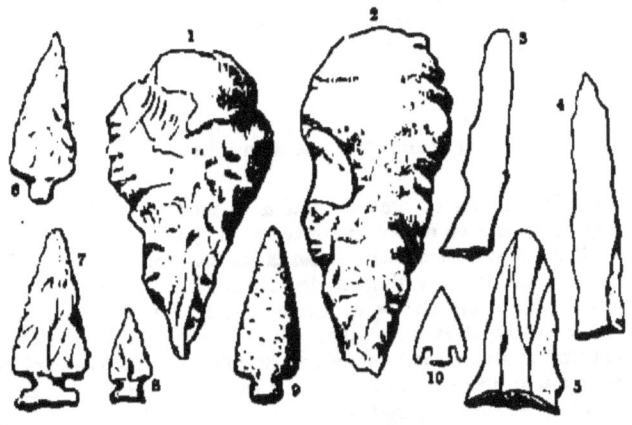

EARLY STONE-IMPLEMENTS.
1, 2. From valley of Somme; 3, 4, 5. England; 6, 7, 8. Canada; 9, 10. Scandinavia.

found in these caverns; and though in some instances man has evidently become the tenant long after the other bones were imbedded, yet in many others the remains have accumulated simultaneously, thus showing that our race was coeval with the mastodon in America, with the elephant in Britain, and with the herds of mammoths that browsed on the ancient river-plains of Siberia. Compared with these events—the era of the mammoth and reindeer and cave-men in western Europe—the human skeletons found in some Continental caverns and osseous breccias, in the river-silts of America, in the peat-bogs of our own island, and in the tufaceous limestone or coral-conglomerate of Guadaloupe, are but of yesterday, and date back at the utmost but a few thousand years.

161. The deposits described in the preceding paragraphs are either of vegetable or of animal origin; but there is an intimate admixture of both in the *soil* or superficial covering of the earth. Strictly speaking, soil is an admixture of decomposed vegetable and animal matter—the decay of plants, and the droppings and exuviæ of animals. Though generally containing a large proportion of earthy ingredients, its dark loamy aspect renders it readily separable from the "subsoil" of sand, clay, or gravel, that lies beneath. It is of universal occurrence, no portion of the earth's crust being uncovered with it, unless, perhaps, the newly-deposited debris on the sea-shore, the shifting sands of the desert, or the snow-clad mountain-top. In some places it barely covers the flinty rock, in others it is several feet in thickness, and everywhere it is annually on the increase.

Igneous or Volcanic Accumulations.

162. The effects of igneous action in modifying the crust of the globe have been already adverted to in para. 24-27, it having been there shown that it acts either in gradual crust-motions—that is, as a gradually elevating or depressing force—as a displacing and deranging force, or as an accumulating agent by discharges of lava, scoriæ, dust, and ashes. Whether manifesting itself in quiet upheavals, in earthquakes, or in volcanoes, its geographical results are of prime importance; and though in certain areas, the present epoch, as compared with some of the past, be one of rest and tranquillity, yet wide regions of the globe bear witness to extensive modifications even within the history of man. Volumes might be filled with the records of such changes; our limits only permit a few recent examples:—Since the commencement of the present century, the shores of the Baltic have been gradually elevated from eighteen to twenty-two inches above their former level, and are still apparently on the uprise. A similar uprise is taking place all along the shores of Siberia, Spitzbergen, and the Arctic islands of North America—the whole being marked by terrace above terrace to the height of several hundred feet. By an earthquake in 1810, a tract in the delta of the Indus, extending to nearly fifty miles in length and sixteen in breadth, was upheaved ten feet, while adjoining districts were depressed, and the features of the delta completely altered. By the great Chili earthquake of 1822, a tract of not less than one hundred thousand square miles was permanently elevated about six feet above its

former level; and part of the sea-bottom remained dry at high-water, with beds of shell-fish adhering to the rocks on which they grew. Within the last ten or twelve years the shores of North Island, New Zealand, have been variously affected, and a large portion in 1856 suddenly but permanently upraised to the height of six or eight feet above its previous level. As with these instances of upheaval so with others of depression, like the western shores of Greenland, the southern shores of Norway, and the seaboard of the Southern States of North America—all of which are said to be undergoing a slow but continued submergence.

163. The above are examples of upheaval and depression on a great scale, and attended with comparatively few convulsions or displacements. The following are of a different order:—In 1692, the town of Port Royal in Jamaica was visited by an earthquake, when the whole island was frightfully convulsed, and about a thousand acres in the vicinity of the town submerged to the depth of fifty feet, burying the inhabitants, their houses, and the shipping in the harbour. The disasters of the great Lisbon earthquake in 1755, when the greater part of that city was destroyed, and sixty thousand persons perished in the course of a few minutes, have been repeatedly recited; as have also those of Calabria, which lasted nearly four years—from 1783 to the end of 1786—producing fissures, ravines, landslips, falls of the sea-cliff, new lakes, and other changes; while the convulsions and disasters which took place in the West Indies, along the Pacific coast of South America, in the Sandwich Islands, in New Zealand, and other volcanic regions during the year 1868, must be fresh in the memory of every newspaper reader.

164. The products of volcanoes, and the effects of volcanic action, have been sufficiently detailed in pars. 25 and 55. The eruptions of Etna and Vesuvius are matters of everyday notoriety; the burying of Herculaneum and Pompeii, a subject of high historic interest. In 1783, the discharges of the Skaptar Jokul, in Iceland, continued for nearly three months, producing the most disastrous effects, as well as most extensive geological changes on the face of the island. "The immediate source, and the actual extent of these torrents of lava, have never been actually determined; but the stream that flowed down the channel of the Skaptar was about fifty miles in length, by twelve or fifteen in its greatest breadth. With regard to its thickness, it was variable, being as much as five hundred or six hundred feet in the narrow channels, but in the plains rarely more than one hundred, and often not exceeding ten feet." Again, the eruption of Mauna Loa (one of the Sandwich Island volcanoes), which took

place in 1855, is described as progressing with amazing force and rapidity, and rolling its wide fiery floods over the mountain's summit down to its base with appalling fury. "Day after day," says an eyewitness, "the action increased, filling the air with smoke, which darkened our entire horizon, and desolating immense tracts, once clothed with waving forests, and adorned with tropical verdure. This eruption has now been in progress for nearly ten months, and still the awful furnace is in blast. The amount of matter disgorged is enormous : the main stream is nearly seventy miles long (including its windings), from one to five miles wide, and varying from ten to several hundred feet in depth." We quote these as instances of hundreds that might be advanced to show the extent of discharges from existing volcanoes. Whether as lava, pumice, scoriæ, dust, hot mud, or ashes, volcanic products, both on land and under the ocean, are materially adding to the structure of the rocky crust, just as in former epochs a similar function was performed by the granites, porphyries, basalts, traps, and trap-tuffs of the mineralogist. Nor is it to the mere accumulation of igneous rock-matter in certain localities that the student must look for the chief results of volcanic effort. As in former epochs, so even now we have lines and axes of volcanic elevation ; and chains of hills, like those pointed out by Von Tschudi in Peru, and by Darwin in the Pacific, have arisen almost within the human era.

RECAPITULATION.

In the preceding chapter we have briefly indicated the nature and extent of the various accumulations that have taken place since the close of the Boulder-drift ; in other words, since sea and land acquired the outlines of their present configuration, and were peopled by existing species. These accumulations we have classed under the head POST-TERTIARY, QUATERNARY, or RECENT, and subdivided into the following groups, according to the agents chiefly concerned in their aggregation :—

FLUVIATILE.—River accumulations and estuary deposits.
LACUSTRINE.—Lake-silts and marls.
MARINE.—Marine silts, sand-drift, shingle-beaches, &c.
ORGANIC.—Peat-mosses, shell-beds, coral-reefs, &c.
CHEMICAL.—Calcareous, silicious, and saline deposits.
IGNEOUS.—Discharges of lava, &c.; earthquake displacements.

As all these agencies are incessantly at work, some of the preceding accumulations are still in progress, others are comparatively recent, and some again of vast extent and antiquity. Indeed, when estuary deposits, alluvium in valleys, lake-silts, peat-mosses, sand-drifts, coral-reefs, igneous discharges, upheavals and submergence of land are taken in the aggregate, they assume a geological importance not at all inferior, as far as amount is concerned, to any of the older formations. Palæontologically, they are also of considerable interest, affording evidence of certain general extinctions, as the mammoth, Irish deer, dinornis, dodo, solitaire, &c.; and of many local removals, as those of the elephant, rhinoceros, wild-boar, elk, bear, wolf, beaver, &c., from the surface of our own islands. In fact, the cosmical conditions of our planet forbid any cessation of progress; and thus, while its inorganic materials are being worn down, shifted, and reconstructed, its vitality must also undergo modifications, redistributions, and it may be extinctions. *Theoretically*, the accumulations of the present era are not only of high interest in themselves, but of prime importance, as furnishing a key to the complicated phenomena of former epochs: *practically*, they present many important features to the farmer, engineer, and navigator; and furnish us industrially with such substances as brick-clays, sand, marl, peat, pumice, pozzolana, sulphur, borax, petroleum, and other similar products.

XVII.

REVIEW OF THE STRATIFIED SYSTEMS.—GENERAL DEDUCTIONS.

165. To present a history of the structure and past conditions of our globe is the object of all geological inquiry. Astronomy may reveal its relations to the other orbs of the planetary system; Geology alone can unfold its individual constitution and structure. At the outset of his inquiry—on the very surface of the rocky crust he is about to examine—the geologist is met by the fact, that everything beneath and around him is in ceaseless action, reaction, and change. The causes of this change he finds in the atmosphere that envelops the earth, in the waters that course and cover its surface, in the life that peoples it, in the chemical constitution of the substances of which it is composed, and in the fires that glow within its interior. These are ever and everywhere active; here wasting and degrading, there accumulating and reconstructing; here submerging the habitable dry land beneath the ocean, there upheaving the sea-bottom to form new islands and continents; and anon preserving in the re-formed material the remains of plants and animals as evidences of the world's geographical conditions at the time of their entombment. As at the present moment, so in all time past similar operations must have been going forward, and the results are manifested in the rock-formations of the solid crust which it is the province of Geology to investigate.

106. In this rocky crust we find sandstones that must have formerly spread out as sandy shores; conglomerates that formed pebbly beaches; shales that were the muds and clays of former lakes and estuaries; limestones that once were living coral-reefs; and coal-beds composed of the remains of a bygone vegetation. Here, also, we discover imbedded corals and shells and fishes that must have lived in the ocean; reptiles that thronged shallow

bays and estuaries; huge mammalia that browsed on river-plains; and plants, some that flourished in the swampy jungle, and others that reared their trunks in the tropical forest. Of all this there is the clearest and most abundant evidence; and by comparing and arranging, by tracing back from the accumulations of yesterday to the deepest-seated strata in the rocky crust, geologists have been enabled to present a pretty vivid outline of the world's history—of all its phases and conditions from the earliest time we have traces of organisation and life, up to the existing order of things, of which Man is the appointed head and ornament.

167. The exponents of this history, we have said, are the rocky strata of the globe; and these, after diligent research in many and distant regions, have been arranged into systems and groups, according to their order of superposition or relative age—each set being spread over certain areas, marked by some peculiarity of mineral composition, and characterised by the remains of certain plants and animals not found in any other series of strata. In fine, each group or formation represents a portion of world-history —the only record of the conditions, the events, and the life of that period being in the nature of the strata themselves, or in the plants and animals they entomb. Tabulated in chronological order, these systems and groups present the following succession: and could we map out their respective areas in the same manner as we do the existing sea and land, and restore the forms of their fossil plants and animals, Geology would have accomplished its task, and done for the past phases of the globe what geography and natural history are doing, and have done, for its present features:—

Systems.	*Groups.*	*Epochs.*
QUATERNARY (POST-TERTIARY).	{ Accumulations in progress. Recent accumulations. }	CAINOZOIC.
TERTIARY.	{ Pleistocene. Pliocene. Miocene. Eocene. }	
CRETACEOUS.	{ Chalk. Greensand. }	
OOLITIC (JURASSIC).	{ Wealden. Oolite. Lias. }	MESOZOIC.
TRIASSIC.	{ Saliferous marls. Muschelkalk. Upper new red sandstone. }	

Systems.	Groups.	Epochs.
PERMIAN.	{ Magnesian limestone. Lower new red sandstone. }	
CARBONIFEROUS.	{ Coal-measures. Millstone grit. Mountain limestone. Lower coal-measures. }	PALÆOZOIC.
OLD RED SANDSTONE (DEVONIAN).	{ Yellow sandstones. Red sandstones and conglomerates. Grey fissile sandstones and pebbly conglomerates. }	
SILURIAN.	{ Upper shales and limestones. Lower grits and flags. }	PALÆOZOIC.
CAMBRIAN.	{ Semi-crystalline slates and schists. }	
LAURENTIAN.	{ Crystalline schists, quartzites, and serpentinous limestones. }	
METAMORPHIC.	Gneissic and granitoid schists.	{ AZOIC, or HYPOZOIC. }

168. Such are the stratified systems composing the crust of the globe, such the types of vegetable and animal life that have successively peopled its surface. In these we find a long gradation of change and progress—not progress from imperfection to perfection, but from humbler to more highly organised forms. From the lowly sea-weeds of the silurian strata and marsh-plants of the old red sandstone, we rise to the prolific club-mosses, reeds, tree-ferns, and gigantic endogens of the coal-measures; from these to the palms, cycads, and pines of the oolite; and from these again to the exogens or true timber-trees of the present era. So also in the animal kingdom; the graptolites and trilobites of the silurian seas are succeeded by the eurypterites and bone-clad fishes of the old red sandstones; these by the sauroid fishes of the coal-measures; the sauroid fishes by the gigantic saurians and reptiles of the oolite; the reptiles of the oolite by the huge mammalia of the tertiary epoch; and these in time give place to present species, with Man as the crowning form of created existence.

169. We have seen that certain agents are ceaselessly modifying the superficial configuration of the earth, and giving rise to new conditions; and as plants and animals are influenced in their forms and distributions by external causes, these new conditions must be accompanied by new phases and arrangements of vitality. So it has been, as we trace graven on the rocky records of geology; so it is; and so, we infer, it will ever continue to be,

under the ceaseless superintendence of an all-wise and beneficent Creator. To discover the facts of this long gradation of change and progress, and to combine the whole into a connected and intelligible history of our planet, is the aim of all geological inquiry—an aim not the less interesting or important that it bears directly on the procuring of those minerals and metals which add so greatly to the comforts and luxuries of life, and contribute so materially to human progress and civilisation. Combining its theoretical interests with its high practical value—the complexity and nicety of its problems as an intellectual exercise with the substantial wealth of its discoveries—the new light it throws on the duration of our planet, and the wonderful variety of its past life, with the certainty it confers on our industrial researches and operations—Geology becomes one of the most important of modern sciences, deserving, indeed, the study of every cultivated mind, and the encouragement of every enlightened government.

GLOSSARIAL INDEX.

⁎ *The figures refer to the sections of the text in which the particular term or subject occurs.*

ACICULAR, needle-shaped (Lat. *acus*, a needle), 55.
Actynolite (Gr. *aktin*, a ray, and *lithos*, a stone), a frequent mineral in granitic compounds, 50; actynolite-schist, 59.
Alluvium, alluvial tracts, 18.
Alum (Lat. *alumen*, Gr. *als*, *alos*, salt), alum-shales of coal-measures, 101; of lias, 125.
Amber (Arab. *ambar*), a fossil tertiary resin, 142.
Amblypterus, carboniferous fish, 95.
Ammonites, of the oolite, 122; of the chalk, 128.
Amorphous (Gr. *a*, without, and *morphe*, regular form), 30, 34.
Amphitherium, a marsupial mammal of the oolite, 122.
Amygdaloid, a frequent variety of trap rock, 50.
Anoplotherium, tertiary mammal, 140.
Anthracite (Gr. *anthrax*, charcoal), a non-bituminous variety of coal, 96; as distinguished from ordinary coal, 85.
Anthracosaurus (Gr. *anthrax*, coal, and *saurus*, lizard), a reptile of the coal period, 95.
Anticline, anticlinal axis, 81.
Apatite, phosphate of lime or phosphorite, found among granitic rocks, 44.
Aqueous agencies, their mode of operation and results, 17.
Archæopteryx (Gr. *archaios*, ancient, and *pteryx*, wing), fossil bird from German oolite, 122.
Archegosaurus (Gr. *archepos*, beginning), reptile of the coal period, 95.
Arenicolites (Lat. *arenicola*, the sand or lob worm), fossil worm tracks or burrows resembling that of the lob-worm, 76.
Asaphus, silurian trilobite, 76.
Asterophyllites, a coal-measure and permian plant, 98.
Astrea (Gr. *astron*, a star), silurian star-coral, 76.
Atmospheric agencies, their operations and results, 16.
Atolls, the name given to coral islands of an annular form, that is, consisting of a circular belt or ring of coral, with an enclosed lagoon, 22; 159.
Augite (Gr. *auge*, lustre), the principal mineral in many trap and volcanic rocks, 36.
Auriferous (Lat. *aurum*, gold, and *fero*, I yield), applied to veins and deposits yielding gold, 72.
Avalanche (Fr. *avalanges*, *lavanches*), an accumulation of ice, or of snow and ice, which descends from precipitous mountains, like the Alps, into the valleys beneath, 14.
Azoic (Gr. *a*, without; *zoe*, life), as applied to stratified rocks, 42, 66.

BACULITES, a chambered shell of the chalk, 128.
Band, a thin continuous stratum, 34.
Basalt, as distinguished from greenstone, 36, 50.
Basin, trough, or syncline of stratified rocks, 81; basins, tertiary, 138.
Beaches, raised or ancient, 27, 155.
Belemnites, of the oolite, 122; of the chalk, 128.
Bellerophon, a multifold shell, characteristic of the mountain limestone, 94.
Berg-mahl (Swedish), mountain-meal, a recent infusorial earth, 21.
Beryx, chalk fish, 129.
Bitumen (Gr. *pitus*, pitch of the pine-tree), as a mineral compound, 86.
Boreal shells from the drift, figured, 136.
Botryoidal (Gr. *botrys*, a cluster of grapes, and *eidos*, form), applied to certain concretionary forms, 108.
Boulder-clay, formation of, 145,147.
Breccia and brecciated (from the Italian), composed of irregular angular fragments, 35, 36.
Bunter (Ger. variegated), a member of the trias, 109.
Burrstone, a silicious rock of the Paris tertiaries, used for millstones, 142.

CAINOZOIC (Gr. *kainos*, recent; *zoe*, life), as applied to fossiliferous strata, 42.

188　　　　GLOSSARIAL INDEX.

Calamites (Lat. *calamus*, a reed), in carboniferous rocks, 98.
Calcaire grossier (Fr., coarse limestone), one of the eocene beds of Paris, 136.
Calcareous springs (Lat. *calx*, *calcis*, lime), 17.
Calc-tuff, calc-sinter, and other calcareous deposits, 158.
Calymene, silurian trilobite, 76.
Cambrian system, 70–73; origin of the term, 70; rocks of the system, 70; fossils of the system, 71; physical aspects of, 72; industrial products of, 73.
Cannel coal, known also as gas and parrot coal, 97.
Carboniferous system (Lat. *carbo*, coal, and *fero*, I yield), description of, 89–101; subdivisions of, 89.
Carpolithes (Gr. *karpos*, fruit, and *lithos*, a stone), fossil fruits of the tertiary beds, 140.
Cataclysm (Gr. *kataklysmos*, an inundation), applied to any violent flood; deluge; debacle, 146.
Catenipora (Lat. *catena*, a chain, and *porus*, a pore or passage), a silurian coral, 76.
Cephalaspis, a peculiar fish of the devonian epoch, 85.
Ceratites (Gr. *keras*, a horn), a kind of ammonite, 110.
Chalk, as distinguished from ordinary limestone, 22; chalk system, 128; fossils, 128.
Chalybeate springs (Gr. *chalybs*, iron), or those impregnated with iron, 17.
Chelrotherium, a fossil batrachian of the trias, 110.
Chemical agencies, their mode of operation and results, 29, 156.
Chemical solution as distinguished from mechanical suspension, 12.
Chert, a peculiar flinty admixture occurring in many limestones, 26.
Chiastolite slate, or schist, a rock of the clay-slate group, 62.
Chili, upheaval of coast, 26.
Chlorite (Gr. *chloros*, greenish), chlorite schist, &c., 36, 59.
Chondrites (*chondrus*, a species of seaweed), marine plants of the chalk, 128.
Cimoliornis, a bird of the chalk, 130.
Classification of rock-systems, 39, 40.
Clay, as distinguished from marl, 36.
Clay-slate, as a rock and group, 62, 65; geographical extent, 64; uses of, 48.
Claystone, an earthy variety of felspar-rock, 30; claystone porphyry, 30.
Cleavage, phenomenon of, 65.
Cleveland, oolitic ironstone of, 125.
Clinkstone, or phonolite (Gr. *phonos*, sound), a trappean rock, 50.
Clymenia, chambered shell of the old red, 85.
Coal, as distinguished from lignite and jet, 30; varieties of, 97.
Coal in old red sandstone, 83; coal-measures, 97; in oolite, 125; in chalk, 133; in tertiary, 142.

Coal-measures, lower, 90; upper, 95.
Coccosteus, a peculiar fish of the devonian epoch, 85.
Columnar and **sub-columnar, defined**, 34.
Confervites, fossil aquatic plants of the chalk epoch, 128.
Conformable and unconformable, 21.
Conglomerate, as distinguished from sandstone, 26; of old red, 81.
Contorted strata, definition of, 21.
Copper-slate, *kupfer-schiefer* of **Germany**, 109.
Coprolites, in coal-measures, 93.
Coral-reefs, nature of, 21, 151.
Corals of silurian system, 75; of mountain limestone, 95; of the oolite, 121; of the chalk, 128; of existing reefs, 152.
Cornifold (*eidos*, like to), coral-like, 91.
Cornbrash, a shelly conglomerate, one of the members of the oolite, said to derive its name from the facility with which it disintegrates and breaks up (brashy) for corn-land, 116.
Cornstones, limestones of the devonian system, 81.
Cosmogony (Gr. *kosmos*, the world, and *gone*, origin) reasonings or speculations respecting the origin of the universe.
Crag (Celt. *craggan*, a shell), a shelly tertiary deposit, found chiefly in Norfolk and Suffolk, 138.
Crag and tail, phenomenon of, 145.
Crater (Gr. *krater*, a cup or bowl), the term applied to the cup-like orifices of volcanoes, 23.
Cretaceous system, account of, 127–133.
Crinoidea (Gr. *krinon*, a lily, and *eidos*, form), lily-like radiata, 76, 96.
Crop, or outcrop of strata, 21.
Crust of the earth, meaning of the term, 2; composition of, 3, 4.
Crust-motions, gradual, 27.
Crystal (Gr. *krystallos*, ice), originally applied to transparent gems, but now used to denote all minerals possessing regular geometrical forms.
Crystalline and sub-crystalline, 35.
Ctenacanthus, fish-spine, 95.
Ctenoid, Ctenoidians, &c., a division of fishes, 84.
Cucumites, tertiary fruits, 140.
Cumbrian and Cambrian strata, proposed arrangement of, 68.
Cumbingstone, fossil footprints at, 110.
Cupressinites (*cupressus*, the cypress), fossils allied to the cypress, 129, 140.
Currents, marine, their influence, 153.
Cyathophyllum (Gr. *kyathos*, a cup, and *phyllum*, leaf), fossil cup-coral, 76.
Cycadites, fossil plants of the trias and oolite, 110, 129.
Cycloid, Cycloidians, &c., a division of fishes, 84.
Cyclopteris, a fern of the coal-measures, 93.
Cystidea (Gr. *kystis*, a bladder), silurian echinoderms, 76.

GLOSSARIAL INDEX. 189

Debris, a convenient term adopted from the French for all heterogeneous accumulations of wasted material, 22.
Degradation (Lat. *de*, down, and *gradus*, a step), the act of wearing or wasting down gradually or step by step, 17.
Deinotherium (Gr. *deinos*, terrible, and *therion*, wild beast), figured, page 164.
Delta, and deltoid, deposits described, 18.
Dendrerpeton (Gr. *dendron*, tree, and *erpeton*, reptile), reptile of the coal period, 95.
Denudation (Lat. *de* down, and *nudus*, naked). The removal of superficial matter, so as to lay bare the inferior strata, is an act of denudation; so also the removal by water of any formation or part of formation.
Detritus (Lat. *de*, from, and *tritus*, rubbed), matter worn or rubbed off rocks by aqueous or glacial action.
Devonian or Old Red Sandstone system, 33-36.
Dicynodon, triassic reptile, figured, 119.
Diluvial drift, Drift, or Boulder formation, 144.
Dip, or inclination, of strata, 21.
Diprotodon, figured, 147.
Dolomite, as a rock, 30; in permian system, 104.
Dromatherium (Gr. *dromatos*, swift-running, and *therion*, wild beast), a mammal of the triassic or permian period, discovered in the red sandstones of Carolina, North America, 119.
Dyke, wall-like masses of igneous matter filling fissures in stratified rocks are so termed, 22.

Earthquakes, their effects as geological agents, 26, 27, 163.
Encrinites, of silurian rocks, 76; of mountain limestone, 96.
Eocene, or lower tertiary group, 136.
Eozoön Canadense, Laurentian fossil, figured, 69.
Erosion (Lat. *erosus*, gnawed or worn away), application of the term, 18.
Escarpment (Fr.), steep bluff edge or precipice, 51.
Estuary deposits, nature of, 151.
Euomphalus (Gr. *eu*, well, and *omphalos*, the navel), a coiled nautiloid shell of the mountain limestones, 95.
Eurypterus (Gr. *eurys*, broad, and *pteron*, a wing or fin), a genus of palæozoic crustaceans, so called from their broad thong-like swimming feet, 76, 83, 94.
Exuviæ (Lat. *exuere*, to cast or throw off). In zoology this term is applied to the moulted or cast-off coverings of animals, such as the skin of the snake, the crust of the crab, &c.; but in geology it has a wider sense, and is applied to all fossil animal remains of whatever description.

Faboideæ, tertiary fruits, 140.
Fault, fissure or dislocation of strata, 33.

Fauna (Lat. *fauna*, rural deities), in geology, 37.
Favosites (Lat. *favus*, a honeycomb), a silurian coral, 76.
Felspar (Ger. *fels*, rock, and *spath*, spar), as a rock, 28; (elspathic traps, 30.
Fens, or marine marshes, 154.
Ferruginous (Lat. *ferrum*, iron), impregnated with iron; *ferriferous*, yielding iron.
Fibrous texture, composed of fibres like asbestos, 30.
Fire-clay of the coal-measures, 101.
Fishes, fossil, chief characteristics of, and Agassiz's classification of, 84.
Fissile structure (Lat. *fissus*, capable of being split), 30.
Flaggy, Flags, and Flagstones, are terms applied to fissile rocks like pavingstone, 24; Arbroath and Caithness flagstones, 35.
Flexures, in stratification, 21.
Flint, as a rock, 26; formation of, in chalk system, 132.
Flora (Lat. *Flora*, the goddess of flowers), in botany, 37.
Fluviatile or river accumulations, post-tertiary, 142.
Foliation, in metamorphic strata, 52.
Foraminifera (Lat. *foramen*, an opening), a class of minute chambered shells, with an orifice in the septa or plates which separate the chambers, 21, 137, 150.
Forests, submerged, 155.
Formations, in geology, 3.
Fossils, fossil remains (*fossus*, dug up), 7.
Freshets or land-floods, their effects, 17.
Fuci, Fucoids, sea-weeds, or focus-like impressions in silurian rocks, 76; in old red sandstone, 83; in oolitic, 120.
Fuller's earth, a variety of absorbent clay, used in the scouring or fulling of woollen cloth, 126, 133.

Ganoid, Ganoidans, &c., a division of fishes, 84.
Garnets, in metamorphic rocks, &c., 61.
Garnetiferous (*fero*, I bear), yielding or containing garnets.
Gault (*provincial*), a member of the chalk system, 124.
Gastrolis, a gigantic bird of the lower tertiary epoch, 140.
Geodes (Gr. *geodes*, earthy), a term applied to rounded pebbles having an internal cavity lined with crystals; also to rounded or nodular pebbles themselves; and to nodules of clay or ironstone, hollow within, or filled with soft earthy ochre, 52.
Geology, object and scope of, 1, 2, 3; theoretical, 9; practical, 10; how to observe, 11.
Geysers, or hot springs of Iceland, 130.
Glacial phenomena of boulder-drift, 146, 147.
Glacier (Lat. *glacies*, ice), the term applied to those masses of ice which ac-

cumulate in the higher gorges and valleys of snow-covered mountains, 16.
Glossopteris, a fern of the oolite, 130; permian, 106.
Gneiss, as distinguished from granite, 66; the gneiss group, 58-66.
Gold, principal repositories of, 72.
Granite, mineral composition of, 36, 46.
Granitic rocks, description of, 42, 42-48; igneous or aqueous origin of, 43; where found, 42; economic uses of, 48.
Graphite (Gr. *grapho*, I write), so called from its use in making writing-pencils. This substance consists almost entirely of pure carbon with a small percentage of iron, the proportions being about 90 to 9. It is also termed plumbago and *black-lead*, from its appearance, though lead does not at all enter into its composition.
Graptolites, characteristic silurian zoophytes, 76.
Gravel, as distinguished from sand and shingle, 3d.
Greensand, a member of the chalk system, 126.
Greenstones, varieties of, 52.
Greywacke (Gr. *grau*, grey, and *wacke*, a mixed clayey rock), 66.
Grit, as distinguished from sandstone — grindstone-grit, millstone-grit, 36.
Group, in geological classification, 40.
Gryphæa (Gr. *grups*, a griffin), a beak-like shell of the oolite, 131.
Gypsum (Gr. *gypsos*, from *ge*, earth, and *epso*, I boil), originally applied to all limestones, 33.
Gyracanthus, fish-spine, 25.

HAMITE, a chambered hook-shaped shell of the chalk, 129.
Harlech grits, lower Cambrian, 70.
Heliolites (Gr. *helios*, the sun, and *lithos*), silurian corals, 76.
Hitch, slip or displacement of strata, 23.
Holoptychius, a fish of the upper old red and lower carboniferous ages, 82, 95.
Hornblende, as a mineral and rock, 36; granitic rock, 45; metamorphic, 59.
Hymenocaris, Cambrian crustacean, figured, 71.
Hypersthene, as a mineral and rock, 36.
Hypozoic (Gr. *hypo*, under, and *zoe*, life), as distinguished from *zoic*, 42, 47.

ICEBERG Gr. *eis*, ice, and *berg*, a mountain), the name given to the mountainous masses of ice often found floating in the arctic and antarctic seas.
Iceland, volcanic discharges in, 184.
Ichnites, or fossil footprints, 110.
Ichthyolite, and ichthyodorulite, fish remains, 81.
Igneous agency, its mode of operation and results, 24-27.
Igneous rocks, nature of, 31; subdivisions of, 43; relations to the stratified rocks, 52.
Indurated, hardened by heat; and in this sense should be kept distinct from "hard" or "compact."
Infusorial accumulations, nature of, 21.
Ironstone, of the coal-measures, 101.

JET, as distinguished from ordinary coal, 86; in lias, 117.
Joints, divisional planes, "backs and cutters," 24.

KAOLIN, a Chinese term for a fine pottery-clay derived from the decomposition of granitic or felspathic rocks, 48.
Keuper and Keuper marls, members of the trias, 109.
Kimmeridge clay, Kimmeridge shale, or "Kim coal," &c., 120.

LABRADOR rocks, upper Laurentian, 68.
Labyrinthodon, a batrachian reptile of the new red sandstone, figured, 110.
Lacustrine or lake deposits, 152.
Laminated (Lat.), composed of thin plates or laminæ; fissile, 34.
Laurentian system, 64-69; origin of the term, 66; rocks of the system, 68; fossils of the system, 69; physical aspects of, 72.
Lava, an Italian term, now universally applied to all molten rock-matter discharged from volcanoes, 23, 36, 54.
Lead, veins of, in mountain limestone, 101.
Leguminosites, fruits of the tertiary epoch, 140.
Lepidodendron, a carboniferous fossil, 28; permian, 106.
Lias, or Liassic strata, 117.
Lignite (Lat. *lignum*, wood), a variety of coal, 36; occuring in oolite, 125; in tertiary beds, 142.
Limestone, as a rock compound, 36; uses of, 101.
Limuloides, carboniferous crustacean, figured, 91.
Lingula, figured, 71, 76; lingula flags, 70.
Lithology, as distinguished from palæontology, 67.
Littoral (Lat. *litus*, the shore), applied to all deposits and operations taking place near or along the shore, in contradistinction to pelagic (*pelagus*, the deep sea) or deep-sea deposits.
Lituites, silurian chambered shell, 76.
Llanberis slates, lower Cambrian, 70.
Llandeilo rocks, the lowest series of silurian strata, 69.
Lodes and veins of mineral matter, 33.
Lossiemouth, reptiliferous sandstones of, 110.
Ludlow rocks, the upper series of silurian strata, 75.

MAGNESIAN limestone, a member of the permian system, 102. Any limestone containing a notable percentage of carbonate of magnesia is termed "magnesian."
Mammoth, or post-tertiary elephant, figured, 160.

GLOSSARIAL INDEX. 191

Manna Loa, eruption of, in 1856, 164.
Marble, as distinguished from limestone, 86; marbles of the metamorphic rocks, 61; of the carboniferous, 63; of the oolite, 125.
Marine deposits, post-tertiary, 153.
Marl, nature and composition of, 86.
Mastodon, a post-tertiary pachyderm, figured, 147.
Mechanical suspension as distinguished from chemical solution, 19.
Megalichthys (Gr. *megals*, great, and *ichthys*, fish), a fish of the coal epoch, 96.
Megalonyx, a gigantic edentate mammal from the upper tertiaries of South America, 147.
Megatherium (Gr. *mega*, great, and *therion*, wild beast), a tertiary mammal, 141.
Mesozoic (Gr. *mesos*, the middle, and *zoe*, life), as applied to fossiliferous strata, 42, 166.
Metals, their mode of occurrence, 36.
Metamorphic system, origin and description of, 57-65.
Mica-schist, as a rock, 86, 53; as a rock-group, 57-61.
Microlestes, a mammal of the triassic period, 110.
Millstone grit, a subdivision of the carboniferous rocks, 96.
Miocene, or middle tertiary group, 135.
Monoclinal strata, 31.
Moraine, a Swiss term for the mounds of detritus (sand, gravel, and boulders) brought down by glaciers, 144.
Mountain limestone, or carboniferous limestone, 93-95.
Murchisonia (after Murchison), a whorled palæozoic univalve, 76.
Muschelkalk (Ger. *muschel*, shell, and *kalk*, lime), a member of the trias, 109.
Mussel-beds, or "Mussel-binds," thin shelly layers occurring in the coal-measures, 96.

Neocomian, a synonyme of the green-sand, 127.
Neozoic (Gr. *neos*, new, and *zoe*, life), as applied to fossiliferous strata, 42.
Nerinæa, a characteristic shell of the oolite, 121.
Neuropteris, a fern of the coal-measures, 98.
New Red Sandstone, subdivisions of, 101.
New Zealand, recent upheaval of coast, 76, 162.
Nipadites, tertiary fruit, 140.
Nummulites, fossil foraminifera of the lower tertiary, 137.

Obsidian (Gr. *opsianus*), a compact vitreous lava or volcanic glass; so called from being polished by the ancients, and used for looking-glasses, 54, 56.
Oceanic currents, effects of, 153.
Ochre, hydrated oxide of iron, as derived from coal-measures, 91.

Odontopteris, carboniferous fern, 98.
Old Red Sandstone system, or Devonian, 79-85; lithology of, 82; fossils of, 83-85; scenery of, 80; where found, 81; uses of, 82.
Oldhamia, figured, 71, 72.
Olenus, Cambrian trilobite, figured, 71.
Oolitic system, description of, 115-126; subdivisions of, 116; oolite group, 116.
Orbitolite limestone of America, 137.
Organic accumulations, of recent growth, 153.
Organic agencies, their mode of operation and results, 20, 152.
Ornithichnites, fossil footprints of birds, 140.
Ornitholites (Gr. *ornis*, bird, and *lithos*, stone), undetermined fossil remains of birds, 140.
Orthoceras, Orthoceratite (Gr. *orthos*, straight, and *keras*, a horn), a genus of straight horn-shaped chambered shells, 76.
Ossiferous caves, gravels, and breccias, 160.
Osteolepis, a fish of the old red sandstone epoch, 85.
Outcrop, of extreme edge of inclined strata, 31.
Outlier, or detached portion of a formation, 31.
Overlying, as applied to overflows of igneous rocks, 32.

Palæonsicus, palæozoic fish, 108.
Palæontology, as distinguished from lithology, 67.
Palæosaurus (Gr. *palaios*, ancient, and *sauros*, a lizard), in permian strata, 106.
Palæotherium, tertiary mammal, 140.
Palæozoic (Gr. *palaios*, ancient, and *zoe*, life), as applied to certain fossiliferous strata, 42.
Palmacites, fossil palms of the oolite, 120.
Paradoxides, Cambrian trilobite, figured, 71.
Peat, peat-moss, formation of, 26, 28, 158.
Pebbles, Scotch, from trap-rocks, 53.
Pecopteris, a fern-like fossil in coal-measures, 98; in permian strata, 106; in oolite, 120.
Pelagic (Gr. *pelagos*, the sea), applied to deep-sea deposits and operations, as distinguished from shore or littoral ones.
Pelagornis (Gr. *pelagos*, the sea; *ornis*, bird), a gigantic tertiary bird, apparently allied to the albatross, 140.
Pentacrinite (Gr. *pente*, five), a five-sided encrinite, 85.
Periclinal strata, 31.
Permian system, described, 102-108; rocks of, 104; fossils of, 105; economic products, 108.
Petaledus, palatal tooth, 84.
Petraphiloides, tertiary fruits, 140.
Petrify, Petrifaction (Lat. *petra*, a stone, and *fio*, I become). All vegetable or

GLOSSARIAL INDEX.

animal matters found in a fossil state, or converted into stony matter, are said to be petrified, 7.

Petroleum (Lat. *petra*, a rock, and *oleum*, oil), literally rock-oil ; a liquid mineral pitch, so called because it seems to come out of the rock like oil, 3d.

Peucites, fossil conifera of the oolite, 120.

Phaneropteuron, old red sandstone fish, 85.

Phascolotherium, oolitic mammal, 122.

Pisolite, or peastone, in oolite, 113.

Pitchstone, and pitchstone porphyry, varieties of igneous rocks, so termed from their pitch-like lustre, 107.

Placoid, Placoidians, &c., a division of fishes, 84.

Plagiaulax, an abbreviation for *Plagiaulacodon* (Gr. *plagios*, oblique ; *aulax*, groove; and *odous*, tooth), an herbivorous marsupial of the oolite, so called from the diagonal groovings of the premolar teeth, 122.

Platemys, marine tertiary turtle, 140.

Platysomus (Gr. *platys*, broad, and *omos*, the shoulder), a ganoid fish of the permian epoch, 108.

Pleistocene, a synonyme of the "Drift," 148; nature and composition of, 143-157.

Plesiosaurus, a saurioid of the oolitic system, 122.

Pliocene, the upper group of the tertiary system, 138.

Poikilitic, a term formerly applied to the new red sandstone, 102.

Polype (Gr. *polys*, many, and *pous*, a foot), the zoological term applied to zoophytes having many tentacula or feet-like organs of prehension ; hence also the term *polypidom*, 22.

Porphyry (Gr. *porphyreos*, purple), originally applied to a reddish igneous rock used in Egyptian architecture, but now applied to all igneous rocks having detached crystals (mostly of felspar) disseminated through the mass ; hence the term *porphyritic*. We have thus porphyritic granites, greenstone porphyries, felspar porphyries, trachytic porphyries, and so forth, 50.

Portland stone, a stratum of the upper oolite, 125.

Post-tertiary, or recent accumulations, 143-164.

Potstone, a soft magnesian rock, the *lapis ollaris* of the ancients, 61.

Primary, Primitive, in geological classification, 38.

Productus, a bivalve characteristic of the carboniferous limestone, 95 ; in magnesian limestone, 102.

Protogine (Gr. *protos*, first, and *ginomai*, I am formed), a granitic rock, 36, 49.

Protosaurus (Gr. *protos*, first, and *sauros*, a lizard), in permian strata, 106.

Psammodus, painted tooth, 95.

Psilophyton, old red sandstone plant, 83.

Pterichthys, fish of the old red sandstone, 84.

Pterodactyle, a flying sauroid of the oolite, 122.

Pterophyllum (Gr. *pteron*, a wing, and *phyllon*, a leaf), fern-like fossils of the oolite, 120.

Pterygotus, a peculiar crustacean of the old red sandstone, 78, 83.

Puddingstone or conglomerate, 36, 44.

Pumice (Ital. *pomice*, allied to *spuma*, froth or scum), a light, porous, froth-like lava, 25, 64, 66.

Puozzolana, uses of, 66.

Pyrites (Gr. *pyr*, fire), sulphurets of iron or of copper, 35; in slate, 62 ; in coal, 101 ; in lias, 117.

QUAQUAVERSAL (Lat., on every side). This term is applied to strata which dip in every direction from a common point or centre of elevation.

Quartz, nature and composition of, 26 ; quartz-rock, or quartzite, 52.

Quaternary, or post-tertiary system, 143-164.

RAGSTONE, applied to coarse concretionary or breccio-concretionary rocks, as coral rag, Kentish rag, &c.

Retepora (Lat. *rete*, a net, and *porus*, a pore), a Flustracea-like zoophyte found in various formations, 25.

Ripple-mark on sandstones, 87.

Rivers, their effects as geological agents, 17, 147.

Rock, geological application of the term, 28 ; description of various rocks, 36 ; classification of, into systems and groups, 37-42.

Rock-salt, deposits of, in England, 110.

Roestone, or oolite, as distinguished from pisolite, 113.

Rhynchosaurus, triassic reptile, 110.

SACCHAROID (*saccharum*, sugar ; *eidos*, like), like loaf-sugar in texture, 35.

Saddle-back, a familiar term for anticlinal strata, 31.

Sal-ammoniac, uses of, 56.

Saliferous (Lat. *sal*, salt, and *fero*, I yield), a term applied to salt-yielding strata, 102.

Salt, Cheshire deposits of, 114 ; lake deposits of, 157.

Salts, combinations of acids and bases, as rock-salt, potash, soda, &c., 35.

Sand, sandstone, descriptions of, 36.

Sand-dunes (Brit. *dune*, a billock), sand-drift, occurrence of, 16, 20.

Sauroids (Gr. *saurus*, a lizard), of the lias, oolite, and wealden, 122.

Scaphite, a chambered boat-shaped shell of the chalk, 129.

Scar, a bluff precipice of rock ; hence "scar limestone" applied to the mountain limestone, as it occurs in the hills of Yorkshire and Westmoreland.

Schist, as distinguished from slate, 52.

Schorl, or black tourmaline, an accidental mineral, in granitic rocks, 47.

Scolite (Gr. *skolios*, *torinoos*), tortuous

GLOSSARIAL INDEX. 193

tube-like worm-burrows that occur in many sandstones, 73.
Scoriæ (Ital. *scoria*, dross), volcanic cinders or cindery-like accumulations, 33, 54.
Seam, restriction of the term, 54.
Secondary, in geological classification, 22.
Section (Lat. *sectus*, cut). The line, actual or ideal, which cuts through any portion of the earth's crust so as to show the internal structure of that portion (just as one would slice a loaf, or saw up a tree), is termed a Section, 22.
Sediment (Lat. *sidere*, to settle down), various kinds, as river, lake, and oceanic, 18.
Sedimentary rocks, description of, 29.
Selenite (Gr. *selene*, the moon, and *ites*, for *lithos*), as distinguished from ordinary gypsum, 30.
Septaria (Lat. *septum*, a division or fence), applied to nodules of ironstone, &c., occurring in the shales of the coal-measures, lias, and other strata, because, when broken up, the interior is often divided into net-like compartments by minute veins of carbonate of lime; 112.
Series, applied to a number of allied strata arranged in sequence, or order of superposition, 41.
Serpentine, a mottled magnesian rock, so called from its serpent-like colours, 36, 43.
Shale (Ger. *schalen*, to peel or shell off), applied to all argillaceous strata that split up or peel off in thin laminæ, 30.
Shell-beds, growth and accumulation of, 21, 132.
Shingle, loose imperfectly rounded stones and pebbles, as distinct from gravel and sand, 29.
Sigillaria, a stem characteristic of the coal epoch.
Silicious springs (Lat. *silex*, flint), or those holding silicious matter in solution, 17.
Silt, fine mud, clay, or sand deposited as a sediment from water, 7, 30.
Silurian system described, 74-79; meaning of the term, 75; rocks of, 75; fossils of, 76; extent of, 77; uses of, 79.
Sinter (Ger. *sintern*, to drop or incrust by dropping), hence, calc-sinter and silicious-sinter, 17, 33.
Skaptar Jokul, eruption of, in 1783, 184.
Slate, as distinguished from shale, 36.
Soil, growth and nature of, 161.
Solfatara (Ital. *solfo*, sulphur), a volcanic fissure or orifice from which sulphureous vapours, hot mud, and steam are emitted.
Spalacotherium, mammal of the oolite, 122.
Spar (Ger. *spath*), a mineralogical term applied to those crystals or minerals which break up into rhombs, cubes, plates, prisms, &c., with smooth cleav-age faces. Hence we have calc-spar felspar, &c.
Sphenopteris, a fern of the coal-measures, 98.
Springs, cold, hot or thermal, and mineral, 17.
Stagonolepis, a crocodilian (?) reptile from the triassic sandstone of Elgin, 85, 119.
Stalactite and Stalagmite (Gr. *stalagma*, a drop), 23.
Steatite (Gr. *stear*, fat), so called from its greasy or soapy feel; soapstone, 36.
Stereognathus, a mammal of the oolite, 122.
Stigmaria, the root of sigillaria, and characteristic of carboniferous rocks, 98.
Stratified rocks, synonymous with aqueous and sedimentary, 30.
Stratum (singular), Strata (plural), meaning of, 3; horizontal, inclined, bent, and vertical, 31; structure of various strata, 34.
Strike, the linear direction of any stratum as it appears at the surface, 31.
Structure of rocks, 34.
Stylonurus, a crustacean of the silurian and old red sandstone, 76, 82.
Sub (Lat. under), often applied in geology to express a less degree of any quality, as sub-columnar, not distinctly columnar; sub-crystalline, indistinctly crystalline; applied also to position, as sub-cretaceous, under the chalk.
Sub-fossil, partly fossil, 144.
Submarine or Submerged forests, 156.
Sulphur, as a volcanic product, 54, 55.
Superposition of rock-formations, 38, 41.
Syenite, a granitic rock, mineral composition of, 38, 40, 59.
Syncline, synclinal axis or trough, 31.
System, limitation of the term, 40, 42.

Talc as a rock, 36; talc-schist, 38.
Talus, the sloping accumulation of debris which takes place at the base of a cliff or precipice exposed to the weathering effects of frost, rains, and other atmospheric agents.
Telerpeton, a reptile of the triassic (?) epoch, 85.
Tertiary, in geological classification, 22.
Tertiary system, described, 132-143.
Tetrapodichnites (Gr. *tetra*, four; *pous*, *podos*, the foot; *ichnos*, a footprint; and *ites*), 119.
Texture of rocks, 34.
Thecodontosaurus (Gr. *theke*, a sheath; *odous*, tooth; and *saurus*, lizard), a permian saurian; so called from the sheath- or dung-in-cone-like structure of its teeth, 119.
Thermal (Gr. *therme*, heat), applied to hot springs and other waters whose temperature exceeds 98° Fahr., 23, 158.

N

GLOSSARIAL INDEX.

Tilestone, any thinly laminated sandstone fit for roofing; applied especially to the flaggy beds at the base of the old red sandstone, 31.
Tilt-up, in stratification, 31.
Trachyte (Gr. *trachys*, rough), a variety of volcanic rock, 56, 132; trachytic traps, 56.
Transition, as applied in classification, 82, 64.
Trap-rocks (Swedish, *trappa*, a stair), nature of, 28; trappean compounds, 49; description of, 49-52; where found, 51; economic uses of, 52.
Travertine, a limestone of modern formation, 144.
Tremadoc slates, upper Cambrian, 72.
Triassic system, described, 102-114.
Trigonia (Gr. *treis*, three, and *gonia*, a corner), fossil bivalve of the oolite and chalk, 121.
Trilobites, characteristic silurian crustacea, 75; in mountain limestone, 25.
Trinucleus, silurian trilobite, 76.
Tripoli, or polishing slate, a flinty infusorial aggregate, 31.
Trough, basin, or syncline of stratified rocks, 81.
Tuff, or tufa (Ital. *tufo*, Gr. *tophos*), originally applied to a light porous lava or pumice; but now applied to all open porous rocks; hence trap-tuff, calctuff, or calcareous tufa, 32.

Unconformable, unconformability among strata, 31.
Unstratified rocks, synonymous with igneous and volcanic, 30.
Upheaval of crust by earthquakes, 24.

Valleys of erosion, 18.
Vancouver Island, coal of, 132.
Vegetable growth and drift, 20.

Veins, in stratified and unstratified rocks, 22.
Vesicular or cellular, like lava, 31.
Volcanic agency, operations and results of, 3, 34, 162.
Volcanic rocks, description and varieties of, 54, 150; where occurring, 55; industrial products of, 56.
Volcanoes (Lat. *Vulcanus*, the god of fire), active, dormant, and extinct, 55.
Vulcanism or Vulcanicity (Lat. *Vulcanus*, the god of fire), a general term for all manifestations of the internal heat of the globe; whether exhibited in hot springs, heated vapours, earthquakes, or volcanoes, 56.

Wacke (Ger.), a term applied to all soft earthy varieties of trap, whether tufaceous or amygdaloidal.
Warp, a local term for marine silt, 134.
Wealden group, or weald strata, described, 119; fossils of, 121, 122.
Weathering, application of the term in geology, 18.
Wenlock rocks, the middle series of silurian strata, 75.
Whinstone, Whin, a Scottish or Saxon designation for greenstone, but by miners applied to almost every hard or indurated rock that comes in their way.

Xiphodon, tertiary mammal, 160.

Zamites, fossil plants of the trias and oolite, 110, 120.
Zechstein (Ger., mine-stone), a term applied in Germany to the magnesian limestone of the permian group, from its containing the *kupfer-schiefer* or copper-slate, which is there worked as an ore of copper, 103.
Zosterites, marine plant, figured, 81.

THE END.

WORKS
BY THE SAME AUTHOR.

———◆———

"Few of our handbooks of popular science can be said to have greater or more decisive merit than those of Mr Page on Geology and Palæontology. They are clear and vigorous in style, they never oppress the reader with a pedantic display of learning, nor overwhelm him with a pompous and superfluous terminology; and they have the happy art of taking him straightway to the face of nature herself, instead of leading him by the tortuous and bewildering paths of technical system and artificial classification." —Saturday Review.

GEOLOGY FOR GENERAL READERS. A Series of
Popular Sketches in Geology and Palæontology. Second Edition, containing several new Chapters. Price 6s.

"This is one of the best of Mr Page's many good books. It is written in a flowing popular style. Without illustration or any extraneous aid, the narrative must prove attractive to any intelligent reader."—*Geological Magazine.*

HANDBOOK OF GEOLOGICAL TERMS, GEOLOGY,
and PHYSICAL GEOGRAPHY. Second Edition, enlarged. 7s. 6d.

INTRODUCTORY TEXT-BOOK OF GEOLOGY.
With Engravings on Wood and Glossarial Index. Eighth Edition. 2s.

"It has not been our good fortune to examine a text-book on science of which we could express an opinion so entirely favourable as we are enabled to do of Mr Page's little work."—*Athenæum.*

ADVANCED TEXT-BOOK OF GEOLOGY, Descriptive
and Industrial. With Engravings and Glossary of Scientific Terms. Fourth Edition, revised and enlarged. 7s. 6d.

"We have carefully read this truly satisfactory book, and do not hesitate to say that it is an excellent compendium of the great facts of Geology, and written in a truthful and philosophic spirit."—*Edinburgh Philosophical Journal.*

"We know of no introduction containing a larger amount of information in the same space, and which we could more cordially recommend to the geological student."—*Athenæum.*

THE GEOLOGICAL EXAMINATOR. A Progressive
Series of Questions adapted to the Introductory and Advanced Text-Books of Geology. Prepared to assist Teachers in framing their Examinations, and Students in testing their own progress and proficiency. Third Edition. 9d.

INTRODUCTORY TEXT-BOOK OF PHYSICAL
GEOGRAPHY. With Sketch-Maps and Illustrations. Third Edition. 2s.

"The divisions of the subject are so clearly defined, the explanations are so lucid, the relations of one portion of the subject to another are so satisfactorily shown, and, above all, the bearings of the allied sciences to Physical Geography are brought out with so much precision, that every reader will feel that difficulties have been removed, and the path of study smoothed before him."—*Athenæum.*

ADVANCED TEXT-BOOK OF PHYSICAL GEO-
GRAPHY. With Engravings. 5s.

"A thoroughly good Text-Book of Physical Geography."—*Saturday Review.*

EXAMINATIONS ON PHYSICAL GEOGRAPHY. A
Progressive Series of Questions adapted to the Introductory and Advanced Text-Books of Physical Geography. 9d.

THE PAST AND PRESENT LIFE OF THE GLOBE.
With numerous Illustrations. Crown 8vo. 6s.

LATELY PUBLISHED.

THE PHYSICAL ATLAS OF NATURAL PHENOMENA.
By ALEX. KEITH JOHNSTON, LL.D., F.R.S.E., &c. A New and Enlarged Edition, consisting of 85 Folio Plates, 27 smaller ones, printed in Colours, with 135 pages of Letterpress and Index. Imperial folio, half-bound morocco, £8, 8s.

"A perfect treasure of compressed information."—*Sir John Herschel.*

SCHOOL ATLAS OF PHYSICAL GEOGRAPHY. Illustrating in a Series of original Designs the elementary facts of Geology, Hydrography, Meteorology, and Natural History. By A. KEITH JOHNSTON, LL.D., &c. A New and Enlarged Edition, brought up to the present time, and including a Hydrographic Map of the British Isles, showing the River-Basins, the Rainfall, and the Elevation of the London Contours; and Two Maps illustrating the Geological Action of Ice and Snow, and a New Geological Map of the British Isles. In all, 20 Coloured Maps, with Descriptive Text. Imp. 8vo, half-bound, 12s. 6d.

SCHOOL ATLAS OF ASTRONOMY. Comprising in 21 Plates a complete Series of Illustrations of the Heavenly Bodies, constructed from Original and Authentic Materials. By A. KEITH JOHNSTON, LL.D., &c. A New and Enlarged Edition, with an Elementary Survey of the Heavens, designed as an accompaniment to this Atlas, by ROBERT GRANT, LL.D., &c., Professor of Astronomy, and Director of the Observatory in the University of Glasgow. 21 Maps, Imp. 8vo, half-bound, 12s. 6d.

A GEOLOGICAL MAP OF EUROPE. By Sir R. I. MURCHISON, D.C.L., F.R.S.; PROFESSOR NICHOL; and A. KEITH JOHNSTON, LL.D., F.R.S.E., &c. 4 feet 2 inches by 3 feet 5 inches. £3, 3s.

A GEOLOGICAL AND PALÆONTOLOGICAL MAP OF THE BRITISH ISLANDS. By PROFESSOR EDWARD FORBES. 21s.

HANDY BOOK OF METEOROLOGY. By Alexander BUCHAN, M.A., Secretary of the Scottish Meteorological Society. Crown 8vo, with eight Coloured Charts and other Engravings. A New and Enlarged Edition. 8s. 6d.

"In all other branches of Meteorology Mr Buchan's book not only retains its previously high position, but in several respects materially surpasses it. The chapter on, and charts of, barometric pressure all over the world, are perhaps the most important contributions to the science since the publication of the kindred work by Dové, 'On the Distribution of Heat,' &c.; and though not so likely to be popular, we shall not be surprised if Buchan's 'Isobars' are not really more important than Dové's 'Isotherms.' It is impossible, without reproducing the beautiful maps, to convey an idea of the course of these lines, but we cannot refrain from pointing out the wide range in the mean annual pressure at the level of the sea, which varies from 30.2 inches near the equator to 29.4 inches in the Arctic and Antarctic circles, thus corroborating Maury's views. High praise must also be accorded to the chapter on temperature in its relation to atmospheric pressure."—*Meteorological Magazine.*

THE ORIGIN OF THE SEASONS; Considered from a Geological Point of View: showing the remarkable disparities that exist between the Physical Geography and Natural Phenomena of the North and South Hemispheres. By SAMUEL MOSSMAN. Crown 8vo. 10s. 6d.

COMPARATIVE GEOGRAPHY. By Carl Ritter.
Translated by W. L. GAGE. 3s. 6d.

WILLIAM BLACKWOOD & SONS, Edinburgh and London.

EDUCATIONAL BOOKS

PUBLISHED BY

WILLIAM BLACKWOOD AND SONS,

EDINBURGH AND LONDON.

―――◆―――

NEW WORKS ON GEOGRAPHY.

BY THE

Rev. ALEXANDER MACKAY,
LL.D. F.R.G.S.

I.

A MANUAL OF MODERN GEOGRAPHY, Mathematical, Physical, and Political. With a copious Index. Crown 8vo, pp. 760, price 7s. 6d.

This volume—the result of many years' unremitting application—is specially adapted for the use of Teachers, Advanced Classes, Candidates for the Civil Service, and proficients in geography generally.

II.

TWELFTH THOUSAND.

ELEMENTS OF MODERN GEOGRAPHY, for the Use of Junior Classes. Crown 8vo, pp. 300, price 3s.

The 'Elements' form a careful condensation of the 'Manual,' the order of arrangement being the same, the river-systems of the globe playing the same conspicuous part, the pronunciation being given, and the results of the latest census being uniformly exhibited. This volume is now extensively introduced into many of the best schools in the kingdom.

III.

TWENTY-EIGHTH THOUSAND.

OUTLINES OF MODERN GEOGRAPHY: A Book for Beginners. 18mo, pp. 112, price 1s.

These 'Outlines'—in many respects an epitome of the 'Elements'—are carefully prepared to meet the wants of beginners. The arrangement is the same as in the Author's larger works. Minute details are avoided, the broad outlines are graphically presented, the accentuation marked, and the most recent changes in political geography exhibited.

IV.

NINETEENTH THOUSAND.

FIRST STEPS IN GEOGRAPHY. 18mo, pp. 56, price 4d. Sewed, or 6d. in Cloth.

V.

GEOGRAPHY OF THE BRITISH EMPIRE. Price 3d.

DR MACKAY'S ELEMENTARY GEOGRAPHIES.

OPINIONS.

Sir RODERICK IMPEY MURCHISON, K.C.B., President of the Royal Geographical Society, in his Anniversary Address, 1864, says of the 'Elements':—"Among the elementary publications, I may direct attention to a useful little work, by the Rev. Alexander Mackay, entitled 'Elements of Modern Geography' (Blackwood and Sons). In a former Address I ventured to commend the 'Manual of Geography' by the same author; and the present production is an improved and careful epitome of that work, which can be recommended as a text-book to be used in the educational establishments of the country. . . . I cannot but admire the assiduity and research displayed in the preparation of this elementary treatise."

A. KEITH JOHNSTON, LL.D. F.R.S.E. F.R.G.S., H.M. Geographer for Scotland, Author of the 'Physical Atlas,' &c. &c.—"There is no work of the kind, in the English or any other language, known to me, which comes so near my *ideal* of perfection in a school-book, on the important subject of which it treats. In arrangement, style, selection of matter, clearness, and thorough accuracy of statement, it is without a rival; and knowing, as I do, the vast amount of labour and research you bestowed on its production, I trust it will be so appreciated as to insure, by an extensive sale, a well-merited reward."

English Journal of Education.—"Of all the Manuals on Geography that have come under our notice, we place the one whose title is given above (the 'Manual') in the first rank. For fulness of information, for knowledge of method in arrangement, for the manner in which the details are handled, we know of no work that can, in these respects, compete with Mr Mackay's Manual."

The London Weekly Review.—"The book (the 'Manual') is a most valuable repertory of the facts of the science, remarkably full and accurate in detail. We cordially and earnestly recommend it for the higher classes in schools, for colleges, and to a permanent place, for the purpose of reference, in the library."

Spectator.—"The best Geography we have ever met with."

Athenæum.—"Full of sound information, including the results of the most recent investigations, such as those of Captain Speke in Africa, and in every respect corresponding to the actual state of geographical knowledge, both physical and political."

Museum.—"We are glad to be able very strongly to commend the 'Elements' to the attention of teachers, as one of the best, one of the very good school-books of geography in existence. We can recommend it on account of its fulness, yet within manageable limits. Its information is the most recent. We have tested its accuracy, by comparison with independent sources of information within our reach, and that in connection with our own country, with Denmark, and the United States: we have in no case found any serious discrepancy. To accuracy and freshness of matter it adds terseness of style and clearness of arrangement,—the latter much aided by varieties of typography."

EDUCATIONAL PUBLICATIONS. 3

IMPROVED EDITIONS.
SCHOOL ATLASES
By A. KEITH JOHNSTON, LL.D., &c.
Author of the "Royal Atlas," the "Physical Atlas," &c.

I.
ATLAS OF GENERAL AND DESCRIPTIVE GEOGRAPHY.
A New and Enlarged Edition, suited to the best Text-Books; with Geographical information brought up to the time of publication. 26 Maps, clearly and uniformly printed in colours, with Index. Imp. 8vo. Half-bd., 12s. 6d.

II.
ATLAS OF PHYSICAL GEOGRAPHY,
Illustrating, in a Series of Original Designs, the Elementary Facts of GEOLOGY, HYDROGRAPHY, METEOROLOGY, and NATURAL HISTORY. A New and Enlarged Edition, containing 4 new Maps and Letterpress. 20 Coloured Maps. Imp. 8vo. Half-bound, 12s. 6d.

III.
ATLAS OF ASTRONOMY.
A New and Enlarged Edition, 21 coloured Plates. With an Elementary Survey of the Heavens, designed as an accompaniment to this Atlas, by ROBERT GRANT, LL.D., &c., Professor of Astronomy and Director of the Observatory in the University of Glasgow. Imp. 8vo. Half-bd., 12s. 6d.

IV.
ATLAS OF CLASSICAL GEOGRAPHY.
A New and Enlarged Edition. Constructed from the best materials, and embodying the results of the most recent investigations, accompanied by a complete INDEX OF PLACES, in which the proper quantities are given by T. HARVEY and E. WORSLEY, MM.A., Oxon. 23 Coloured Maps. Imp. 8vo. Half-bd., 12s. 6d.
"This edition is so much enlarged and improved as to be virtually a new work, surpassing everything else of the kind extant, both in utility and beauty." —*Athenæum.*

V.
ELEMENTARY ATLAS OF GENERAL AND DESCRIPTIVE GEOGRAPHY,
For the use of Junior Classes; including a MAP OF CANAAN and PALESTINE, with GENERAL INDEX. 8vo, half-bd., 5s.
"The plan of these Atlases is admirable, and the excellence of the plan is rivalled by the beauty of the execution. . . . The best security of the accuracy and substantial value of a School Atlas is to have it from the hands of a man like our Author, who has perfected his skill by the execution of much larger works, and gained a character which he will be careful not to jeopardise by attaching his name to anything that is crude, slovenly, or superficial."—*Scotsman.*

NEW ATLAS by A. KEITH JOHNSTON.

THE HANDY ROYAL ATLAS.
By ALEX. KEITH JOHNSTON, LL.D. &c.
Author of the 'Royal Atlas,' the 'Physical Atlas,' &c.

45 MAPS CLEARLY PRINTED AND CAREFULLY COLOURED, WITH GENERAL INDEX.

Imperial Quarto, price £2, 12s. 6d., half-bound morocco.

This work has been constructed for the purpose of placing in the hands of the public a useful and thoroughly accurate ATLAS of Maps of Modern Geography, in a convenient form, and at a moderate price. It is based on the 'Royal Atlas,' by the same author; and, in so far as the scale permits, it comprises many of the excellences which its prototype is acknowledged to possess. The aim has been to make the book strictly what its name implies, a Handy Atlas — a valuable substitute for the 'Royal,' where that is too bulky or too expensive to find a place, a needful auxiliary to the junior branches of families, and a *vade mecum* to the tutor and the pupil-teacher.

"Is probably the best work of the kind now published."—*Times*.

"Not only are the present territorial adjustments duly registered in all these maps, but the latest discoveries in Central Asia, in Africa, and America, have been delineated with laborious fidelity. Indeed, the ample illustration of recent discovery, and of the great groups of dependencies on the British Crown, renders Dr Johnston's the best of all Atlases for English use."—*Pall Mall Gazette*.

"This is Mr Keith Johnston's admirable Royal Atlas diminished in bulk and scale, so as to be, perhaps, fairly entitled to the name of 'handy,' but still not so much diminished but what it constitutes an accurate and useful general Atlas for ordinary households."—*Spectator*.

"He has given us in a portable form geography posted to the last discovery and the last Revolution."—*Saturday Review*.

Fourth Edition, 1s. 6d.

ENGLISH PROSE COMPOSITION,
A PRACTICAL MANUAL FOR USE IN SCHOOLS.

BY
JAMES CURRIE, M.A.,
PRINCIPAL OF THE CHURCH OF SCOTLAND TRAINING COLLEGE, EDINBURGH.

"We do not remember having seen a work so completely to our mind as this, which combines sound theory with judicious practice. Proceeding step by step, it advances from the formation of the shortest sentences to the composition of complete essays, the pupil being everywhere furnished with all needful assistance in the way of models and hints. Nobody can work through such a book as this without thoroughly understanding the structure of sentences, and acquiring facility in arranging and expressing his thoughts appropriately. It ought to be extensively used."—*Athenæum, September 21, 1867*.

WORKS
ON
GEOLOGY AND PHYSICAL GEOGRAPHY.
By DAVID PAGE, LL.D. F.R.S.E. F.G.S.

GEOLOGY FOR GENERAL READERS. A Series of Popular Sketches in Geology and Palæontology. Second Edition, containing several new Chapters. Price 6s.
"This is one of the best of Mr Page's many good books. It is written in a flowing popular style. Without illustration or any extraneous aid, the narrative must prove attractive to any intelligent reader."—*Geological Magazine.*

HANDBOOK OF GEOLOGICAL TERMS, GEOLOGY, AND PHYSICAL GEOGRAPHY. Second Edition, enlarged. 7s. 6d.

INTRODUCTORY TEXT-BOOK OF GEOLOGY. With Engravings on Wood and Glossarial Index. Seventh Edition. 2s.
"Of late it has not been our good fortune to examine a text-book on science of which we could express an opinion so entirely favourable as we are enabled to do of Mr Page's little work."—*Athenæum.*

ADVANCED TEXT-BOOK OF GEOLOGY, Descriptive and Industrial. With Engravings, and Glossary of Scientific Terms. Fourth Edition, revised and enlarged. 7s. 6d.
"We have carefully read this truly satisfactory book, and do not hesitate to say that it is an excellent compendium of the great facts of Geology, and written in a truthful and philosophic spirit."—*Edinburgh Philosophical Journal.*
"We know of no Introduction containing a larger amount of information in the same space, and which we could more cordially recommend to the geological student."—*Athenæum.*

THE GEOLOGICAL EXAMINATOR. A Progressive Series of Questions, adapted to the Introductory and Advanced Text-Books of Geology. Prepared to assist Teachers in framing their Examinations, and Students in testing their own Progress and Proficiency. Third Edition. 9d.

INTRODUCTORY TEXT-BOOK OF PHYSICAL GEOGRAPHY. With Sketch-Maps and Illustrations. Third Edition. 2s.
"A work which cannot fail to be useful to all who are entering on the study of Physical Geography. We believe, indeed, that many will be induced to enter on the study from a perusal of this little work. The divisions of the subject are so clearly defined, the explanations are so lucid, the relations of one portion of the subject to another are so satisfactorily shown, and, above all, the bearings of the allied sciences to Physical Geography are brought out with so much precision, that every reader will feel that difficulties have been removed, and the path of study smoothed before him."—*Athenæum.*

ADVANCED TEXT-BOOK OF PHYSICAL GEOGRAPHY. With Engravings. 5s.
"A thoroughly good Text-Book of Physical Geography."—*Saturday Review.*

EXAMINATIONS ON PHYSICAL GEOGRAPHY. A Progressive Series of Questions, adapted to the Introductory and Advanced Text-Books of Physical Geography. 9d.

THE PAST AND PRESENT LIFE OF THE GLOBE. With numerous Illustrations. Crown 8vo, 6s.

CHIPS AND CHAPTERS. A Book for Amateurs and Young Geologists. 5s.

**FORTIFICATION: FOR OFFICERS OF THE ARMY AND STU-
DENTS OF MILITARY HISTORY.** By Lieut. HENRY YULE, Bengal
Engineers. With Illustrations. 8vo, 10s. 6d.

"An excellent manual; one of the best works of its class."—*British Army
Despatch.*

In post 8vo, price 5s.

ELEMENTARY ARITHMETIC. By Edward Sang, F.R.S.E.

This treatise is intended to supply the great desideratum of an intellectual
instead of a routine course of instruction in Arithmetic.

In crown 8vo, price 5s.

THE HIGHER ARITHMETIC. By the same Author. Being a
Sequel to 'Elementary Arithmetic.'

"We know, indeed, of no more complete philosophy of pure arithmetic than
they contain; they are well worthy of Sir John Leslie's favourite pupil. It is
almost needless to add, that we consider the reasoning of these volumes both
thorough and close, and the expression of that reasoning uniformly simple and
clear."—*Edinburgh Weekly Review.*

Price Sixpence, for the Waistcoat Pocket.

FIVE PLACE LOGARITHMS. Arranged by E. Sang, F.R.S.E.

TREATISE ON ARITHMETIC, with numerous Exercises for Teaching
In Classes. By JAMES WATSON, one of the Masters of Heriot's Hospital.
Foolscap, 1s.

AINSLIE'S LAND-SURVEYING. A New and Enlarged Edition, embracing RAILWAY, MILITARY, MARINE, and GEODETICAL SURVEYING. By
W. GALBRAITH, M.A. In 8vo, with plates in 4to, price 21s.

"The best book on Land-Surveying with which I am acquainted."—WM.
RUTHERFORD, LL.D. F.R.A.S., *Royal Military Academy, Woolwich.*

SIR WILLIAM HAMILTON'S LECTURES ON METAPHYSICS.
Edited by the Rev. H. L. MANSEL, B.D. LL.D., Waynflete Professor of
Moral and Metaphysical Philosophy, Oxford; and JOHN VEITCH, M.A.,
Professor of Logic and Rhetoric in the University of Glasgow. Fourth Edi-
tion. In 2 vols. 8vo, price 24s.

SIR WILLIAM HAMILTON'S LECTURES ON LOGIC. Edited by
Professors MANSEL and VEITCH. Second Edition. In 2 vols. 8vo, price 24s.

**INSTITUTES OF METAPHYSIC. The Theory of Knowing and
Being.** By JAMES F. FERRIER, D.A. Oxon., late Professor of Moral Philosophy and Political Economy, St Andrews. Second Edition. Crown 8vo,
price 10s. 6d.

DESCARTES ON THE METHOD OF RIGHTLY CONDUCTING THE
REASON, and Seeking Truth in the Sciences; and his MEDITATIONS, and
SELECTIONS from his PRINCIPLES OF PHILOSOPHY. In one vol.,
price 4s. 6d.

CHOIX DES MEILLEURES SCENES DE MOLIÈRE, avec des Notes
de Divers Commentateurs, et autres Notes Explicatives. Par Dr E. DEREC.
Fcap. 8vo, price 4s. 6d.

EDUCATIONAL PUBLICATIONS.

Sixteenth Edition.

EPITOME OF ALISON'S HISTORY OF EUROPE, for the Use of Schools and Young Persons. Post 8vo, pp. 604, price 7s. 6d. bound in leather.

In compiling this Epitome, it has been specially held in view to omit or suppress no *fact* of the slightest importance, and to limit the abridgment to the condensation of the minor and accessory details; and it is trusted that an adherence to this rule, while it has produced a work in which the interest of the narrative never flags, has also secured a history of the time in all essential particulars as complete as the more voluminous records of it.

A Chronological Table has been added of all the principal events, so arranged as to give a clear idea of the order in which they succeed each other; and a full Table of Contents, containing a synopsis of the subjects treated of in the body of the work.

ATLAS to Epitome of the History of Europe. Eleven Coloured Maps, by A. Keith Johnston, LL.D. F.R.S.E. In 4to, price 7s.

School Edition, post 8vo, with Index, price 6s.

HISTORY OF FRANCE, from the Earliest Times to 1848. By the Rev. James White, Author of 'The Eighteen Christian Centuries.'

"This book makes an attempt to furnish a readable account of the country with which we are in closest neighbourhood, and yet of whose history the generality of us know less than of that of almost any other kingdom. It aims at something higher than a mere epitome, for it founds itself on a great deal of various reading, and gives results more than abstracts. At the same time it devotes sufficient space to any occurrences which seem to have a general bearing on the progress or character of the nation. But it does not profess to be very minute in its record of trifling or uninfluential occurrences, nor philosophic in searching out the causes of obscure events."—*Author's Preface.*

"Contains every leading incident worth the telling, and abounds in word-painting, whereof a paragraph has often as much active life in it as one of those inch-square etchings of the great Callot."—*Athenæum.*

School Edition, post 8vo, with Index, price 6s.

THE EIGHTEEN CHRISTIAN CENTURIES. By the Rev. James White, Author of 'The History of France.'

"He has seized the salient points—indeed, the governing incidents—in each century, and shown their received bearing as well on their own age as on the progress of the world. Vigorously and briefly, often by a single touch, has he marked the traits of leading men; when needful, he touches slightly their biographical career. The state of the country and of society, of arts and learning, and, more than all, of the modes of living, are graphically sketched, and upon the whole with more fulness than any other division."—*Spectator.*

"By far the best historical epitome we have ever perused, and it supplies a great want in this knowing age."—*Atlas.*

ELEMENTARY TEXT-BOOK OF SCRIPTURE HISTORY. By Thomas Struthers. Part I. From the Creation to the Death of Moses. Price 6d.

A GLOSSARY OF NAVIGATION. Containing the Definitions and Propositions of the Science, Explanation of Terms, and Description of Instruments. By the Rev. J. D. Harbord, M.A., St John's College, Cambridge; Chaplain and Naval Instructor, R.N. In Crown Octavo, Illustrated with Diagrams, price 6s.

DEFINITIONS AND DIAGRAMS IN ASTRONOMY AND NAVIGATION. By the Same. Price 1s. 6d.

COMPARATIVE GEOGRAPHY. By Carl Ritter. Translated by W. L. Gage. Fcap., price 3s. 6d.

Just published, a New Edition, price 8s. 6d. cloth, with Engravings and Charts.

A HANDY BOOK
OF
METEOROLOGY.
BY
ALEXANDER BUCHAN, M.A.,
Secretary of the Scottish Meteorological Society.

EXTRACTS FROM REVIEWS OF FIRST EDITION.

"A very handy book this, for in its small compass Mr Buchan has stored more and later information than exists in any volume with which we are acquainted."—*Symons's Monthly Meteorological Magazine.*

"To those who wish to have a really 'handy' book on meteorology, clear, concise, and easy of reference, we should certainly recommend Mr Buchan's work."—*The Field.*

"We know of no modern English treatise on meteorology that can compare, in comprehensiveness and conciseness, originality and accuracy, with Mr Buchan's unpretending little manual."—*Nonconformist.*

"After a minute examination of Mr Buchan's book, we feel entitled to say that it is really the best and *handiest* book on the subject that we know; and we have studied a good many of them."—*Aberdeen Journal.*

"Admirably fitted for the object which it is designed to serve—that of assisting people who have not made scientific matters their especial study to form intelligent notions on the subject of which it treats."—*Weekly News.*

"A volume such as this, which explains not only all meteorological phenomena, so far as they are at present understood, but the method of using the various instruments required for the purpose of taking observations, and of repairing them when out of order, cannot fail to be widely appreciated."—*Weekly Despatch.*

"It is also well suited to be used as a text-book in educational establishments, where, we hope, the study of meteorology will be introduced in common with that of the kindred sciences."—*Farmer.*

"We do not know a better book on meteorology; and certainly it is by far the best book on that subject which we know for the horticulturist, the farmer, and, speaking generally, the unscientific reader. Every gardener of the least pretensions to an intelligent knowledge of his profession should not only possess it, but thoroughly master it."—*Gardeners' Chronicle.*

"We have placed it in the row of authorities on our table, ready for reference; for it is, most truly, what it is designated, 'A Handy Book.' It is one of those books, too few in number, which contain nothing but what is desirable to be in its pages, and all is told clearly and pleasantly, as no one can narrate except a writer who is thoroughly master of his subject. We have not often the pleasure of speaking thus of a publication, and every reader of the volume will assent to our opinion of its merits."—*Journal of Horticulture.*

www.ingramcontent.com/pod-product-compliance
Lightning Source LLC
Chambersburg PA
CBHW020923230426
43666CB00008B/1543